Learning
From
NAEP

This material is based on work supported by the National Science Foundation through grant number ESI 0138733 to the National Council of Teachers of Mathematics. Any opinions, findings, and conclusions or recommendations expressed in this material are those of the authors and do not necessarily reflect the views of the National Science Foundation.

Learning
From
NAEP
Professional Development Materials for Teachers of Mathematics

Edited by
Catherine A. Brown
Lynn V. Clark

NATIONAL COUNCIL OF
TEACHERS OF MATHEMATICS

Copyright © 2006 by
THE NATIONAL COUNCIL OF TEACHERS OF MATHEMATICS, INC.
1906 Association Drive, Reston, VA 20191-1502
(703) 620-9840; (800) 235-7566; www.nctm.org

Library of Congress Cataloging-in-Publication Data

Learning from NAEP : professional development materials for teachers of
mathematics / edited by Catherine A. Brown, Lynn V. Clark.
    p. cm.
    Includes bibliographical references.
    ISBN 0-87353-590-1
    1. Mathematics teachers--Training of--United States. 2. Mathematics—
Study and teaching—Standards—United States. 3. National Assessment of
Educational Progress (Project) I. Brown, C. A. (Catherine Anne), 1969-
II. Clark, Lynn V.
    QA10.5.L43 2006
    510.71'073--dc22

                                                          2006010257

The National Council of Teachers of Mathematics is a public voice of mathematics education,
providing vision, leadership, and professional development to support teachers
in ensuring mathematics learning of the highest quality for all students.

PRINTED IN THE UNITED STATES OF AMERICA

Contents

Contents

Contents of the CD-ROM

Learning From NAEP Home

Workshops and Activity Bank

        Student Understanding (chapter 5)

        Mathematical Content (chapter 6)

        Assessment (chapter 7)

        State NAEP (chapter 8)

        Equity (chapter 9)

        Exploring Graph Sense (chapter 10)

NAEP Item Search

Resources

        The Nation's Report Card

        Activity Bank

        References

        Glossary

        Monograph Excerpts

About the IU-NAEP Project

        Meet the Learning From NAEP Group

        Monograph Excerpts

In today's educational climate of accountability based on high-stakes tests, many educators are eager for guidance in preparing their students and understanding more about how performance on these tests relates to student learning. Although the National Assessment of Educational Progress (NAEP) is not currently considered a high-stakes test, its results provide information about overall student performance as well as performance related to specific content strands, test items, and grade levels. Those results can be used to help educators make decisions about classroom instruction and implement evidence-based practices that will best meet the needs of their students.

Developing the *Learning From NAEP* Materials

The *Learning From NAEP* professional development materials are part of a larger project, IU-NAEP, a collaborative effort of the NCTM and mathematics educators at Indiana University with financial support from the National Science Foundation. The IU-NAEP project has two primary foci. Focus I involves the preparation of interpretive reports of data from various NAEP mathematics assessments and has produced two monographs: *Results and Interpretations of the 1990 Through 2000 Mathematics Assessments of the National Assessment of Educational Progress,* edited by Peter Kloosterman and Frank K. Lester, Jr. (Reston, VA: National Council of Teachers of Mathematics, 2004) and *Results and Interpretations of the 2003 Mathematics Assessment of the National Assessment of Educational Progress,* edited by Peter Kloosterman and Frank K. Lester, Jr. (Reston, VA: National Council of Teachers of Mathematics, forthcoming). Focus II has involved the creation of a set of materials designed to help educators better understand the intricacies of assessment data and how such data relate to student learning in mathematics classrooms. The *Learning From NAEP* manual and its accompanying CD-ROM are the result of this second focus.

Creating Quality Professional Development for Mathematics Educators

A need exists for high-quality professional development materials that bring together the most recent data on mathematics achievement, a variety of tools to access those data and activities, and background materials to help educators understand the implications of those data. *Learning From NAEP* builds a bridge between the world of research and the classroom, drawing on the experiences of educators and the expertise of researchers to create a set of research-based materials that address issues that are meaningful to classroom teachers. A special effort has been made by the authors to highlight the words of researchers throughout this manual.

GLOSSARY

High-stakes test — An assessment instrument used to make significant educational decisions about students, teachers, schools, or school districts

Item — The basic, scorable part of an assessment; a test question

NAEP content strands — Content areas used in the Main and State mathematics NAEPs: (a) Number Sense, Properties, and Operations; (b) Measurement, (c) Geometry and Spatial Sense; (d) Data Analysis, Statistics, and Probability; and (e) Algebra and Functions

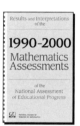

Look for the monograph icon on the CD-ROM and in the manual to see quotes and resources from the monograph (Kloosterman & Lester, 2004).

Visit the Resources section of the CD-ROM to access a downloadable list of references for each chapter.

RESEARCH

"A true profession of teaching will emerge as teachers find ways and are given the opportunities to improve teaching. By improving teaching, we mean a relentless process in which teachers do not just improve their own skills but also contribute to the improvement of Teaching with a capital T. Only when teachers are allowed to see themselves as members of a group, collectively and directly improving their professional practice by improving pedagogy and curricula and by improving students' opportunities to learn, will we be on the road to developing a true profession of teaching." (Stigler, J.W., & Hiebert, J. [1999]. *The teaching gap: Best ideas from the world's teachers for improving education in the classroom.* New York: Free Press.)

At the beginning of each workshop, you will find a checklist of the Principles and Process Standards that are emphasized in that workshop. The corresponding NAEP content strands, which align with the NCTM Content Standards, can be found above the materials checklist, and may serve as a guide as you modify the workshop on the CD-ROM.

Designing high-quality professional development for mathematics educators requires close attention to many facets of teacher learning. Teachers, like students, learn best with a coherent curriculum. They also learn best in the types of learner-centered environments that are conducive for actively constructing new knowledge about teaching and learning mathematics, be that learning transformative—requiring large shifts in practice, beliefs, and knowledge—or additive—adding skills to an existing repertoire (Thompson & Zeuli, 1999).

Principles and Standards for School Mathematics (Principles and Standards) (NCTM, 2000) provides a theoretical framework for these professional development materials. *Principles and Standards* has three major components. First, the Principles for school mathematics, which reflect basic perspectives on which educators should base decisions that affect school mathematics. These Principles establish a foundation for school mathematics programs by considering the broad issues of equity, curriculum, teaching, learning, assessment, and technology; they also guided the focus of the professional development workshops included here and on the accompanying CD.

The Content and Process Standards for school mathematics describe an ambitious and comprehensive set of goals for mathematics instruction. The five Content Standards present goals in the mathematical content areas of number and operations, algebra, geometry, measurement, and data analysis and probability, which are also reflected in the NAEP content strands. The five Process Standards describe the goals for the processes of problem solving, reasoning and proof, connections, communication, and representation. Together, the Standards describe the basic skills and understanding that students will need to function effectively in the twenty-first century. Care has been taken by the authors to weave into the workshop activities opportunities to examine how student responses represent the Process Standards.

No one activity or workshop can address everyone's needs. However, individual professional development activities and workshops can be interwoven with other opportunities for participants to collaborate to form a comprehensive and cohesive professional development experience. For strategies to accomplish this goal, see *Designing Professional Development for Teaching of Science and Mathematics* (Loucks-Horsey, S., Love, N., Stiles, K.E., Mundry, S., & Hewson, P.W. [2003]. Thousand Oaks, CA: Corwin Press).

The Writing Group

The *Learning From NAEP* writing team draws on the expertise of Indiana University faculty members, Indiana University graduate students, and former Indiana University graduate students who now teach at other institutions. All have a deep respect for teachers and enjoy working with them in a variety of settings. When the funding for the IU-NAEP project was awarded, this team of authors, collectively known as the Learning From NAEP Group, was brought together to create a set of professional development materials based on released NAEP test items, available student responses, NAEP achievement data, and Focus I interpretive analyses. In large part, the catalyst for these professional development materials was a collection of more than 5,000 student responses to NAEP test items. Because NAEP is administered to thousands of students each year, the pool of student responses yields some very "interesting" examples of student understanding. Although the student responses greatly extended the resources currently available on the NAEP Web site, the student work was not enough. The writing group soon discovered that the student responses became compelling in the context of the Focus I meta-analysis, and the group was able to use those responses to address the larger issues already being examined in the monograph.

Because NAEP is administered to thousands of students each year, the pool of student responses yields some very "interesting" examples of student understanding.

Next the writing group worked for over a year to assemble a databank of classroom-tested activities that would provide entry points to examine and interpret the data and student responses. Those activities became the foundation for the workshops in this manual and the Activity Bank on the accompanying CD-ROM. In addition, the authors worked with the CD-ROM designers to create tools that would provide access to the larger set of student work and supporting data (e.g., the NAEP Item Search). Finally, the Learning From NAEP Group field-tested the materials and submitted them for national review.

Please note that the professional development activities and workshops contained in this book do not constitute a comprehensive, coherent professional development experience for mathematics educators. Instead, our hope is that the activities and workshops in this manual, and the variations you can create through the use of the accompanying CD, can be woven into various programs of high-quality professional development.

GLOSSARY

NAEP Web site (online tools) — Maintained by the National Center for Educational Statistics, www.nces.ed.gov/nationsreportcard/ lists NAEP items, student performance, questionnaire results, and state-specific data.

 A list of 14 generic activities is available in the Resources section of the CD-ROM Activity Bank. These activities were drawn from the collected experience of the writing group and served as the building blocks for the different workshops in this manual. A description of each activity is also available in Appendix A of this manual.

RESEARCH

"In my judgment, these user friendly professional development materials have the potential to assist K–12 teachers in evaluating programs, analyzing data and student misconceptions, and ultimately improving instruction. The materials use NAEP resources as a means to develop teachers' analytical and diagnostic skills, as well as an awareness of key issues in mathematics learning and teaching. The professional development strategies embedded in the materials are sound and represent an approach that engages teachers and administrators in reflection and analysis."
—William S. Bush, Director,
Center for Research in
Mathematics and
Science Teacher Development

Acknowledgments

The following writers and developers make up the Learning from NAEP Group. The group members spent long hours both together and apart to collaboratively create these professional development materials. The process of creation and the resulting manual and companion CD were made exciting by the diverse perspectives and talents of the members of the Learning From NAEP Group

Workshop and Activity Authors

Catherine A. Brown, chair
Indiana University, Bloomington, Indiana

Fran Arbaugh
University of Missouri, Columbia, Missouri

Beatriz D'Ambrosio
Miami University, Oxford, Ohio

Myoungwhon Jung
Indiana University, Bloomington, Indiana

Signe E. Kastberg
Indiana University–Purdue University Indianapolis, Indianapolis, Indiana

YoungOk Kim
Indiana University, Bloomington, Indiana

Diana V. Lambdin
Indiana University, Bloomington, Indiana

Kathleen Lynch-Davis
Appalachian State University, Boone, North Carolina

Rebecca McGraw
University of Arizona, Tucson, Arizona

Shelby P. Morge
Indiana University, Bloomington, Indiana

Christine Oster
Childs Elementary School, Bloomington, Indiana

Paula R. Stickles
Indiana University, Bloomington, Indiana

Crystal Walcott
Indiana University, Bloomington, Indiana

Indiana University Design and Development Team

Lynn V. Clark, chair

Catherine A. Brown

Myoungwhon Jung

YoungOk Kim

Shelby P. Morge

Christine Oster

Nathan Walton, CD-ROM designer

Jeff Hanson, CD-ROM designer

The editors would like to thank Peter Kloosterman and Frank Lester for their thoughtful contributions to these materials.

The editors also wish to thank the following reviewers for their advice and insights during the development of these materials:

William S. Bush
Professor, Mathematics Education, University of Louisville
Director, Center for Research in Mathematics and Science Teacher Development

Mary M. Lindquist
Emeritus Callaway Professor of Mathematics Education, Columbus State University
Past-president of the National Council of Teachers of Mathematics

Marilyn Strutchens
Associate Professor, Auburn University

Introduction: About This Book

The *Learning From NAEP* professional development materials included in this manual and the accompanying CD-ROM provide opportunities for educators to examine data from the National Assessment of Educational Progress (NAEP) within the context of research-based and field-tested workshops. The background in this manual, activities in the workshops, and resources and materials on the CD-ROM furnish educators with ways to navigate assessment issues and data as they move toward evidence-based practices and policy.

> *The Learning from NAEP professional development materials are intended to empower educators to use standardized assessment data to better understand student learning and to improve teaching.*

Navigating the *Learning From NAEP* Materials

The first four chapters of the manual present background on NAEP (chapter 1), the kinds of data it generates (chapter 2), and the tools you will need to access these data (chapters 3 and 4). Each of the next five chapters is a workshop that addresses a particular theme related to the analysis of data from NAEP mathematics examinations: student understanding, content knowledge, assessment, state issues, and equity. Each workshop can be modified: some by NAEP content strand, others by state or grade level (4th, 8th, and 12th). Chapter 10 deconstructs the process of adapting a workshop and demonstrates how to use the resources on the CD-ROM to personalize a workshop.

Facilitating the Workshops

The manual includes several guideposts to help you facilitate the workshops described in chapters 5 through 10. Each workshop chapter begins with an overview, followed by a checklist of the ways in which the workshop can be modified (e.g., grade level and content strand). This section is followed by a table that lists all the materials needed and gives reference numbers for those materials available for download on the CD-ROM. In addition, the authors provide "Background and Context Notes" that often include research and tips for "Preparing to Facilitate." The workshop is divided into activities and steps. The "Facilitating the Workshop" section describes each activity in detail and makes suggestions for implementation.

The facilitator is crucial to the success of these workshops. The information included in the first four chapters of this manual can be supplemented by additional readings listed in the References section of the

RESEARCH

The facilitator's role "is very important in ensuring that the classroom norms are supportive of [participants'] learning in this way and in pressing [participants] to think deeply about their solution methods and those of [the other participants] and, most important, about the mathematics they are learning" (Lester & Charles 2003).

To help facilitate the workshops, each workshop has a Facilitator's PowerPoint that can be found on the Workshop Materials page on the accompanying CD-ROM.

CD-ROM and in Kloosterman & Lester (2004). Chapter 10 fosters insight into the dynamics of facilitating a workshop and how the facilitator can encourage collaboration among participants. Most important, the facilitator sets the tone of investigation rather than of the evaluation of student responses. The student responses provided on the accompanying CD-ROM are intended for the purpose of analysis, not as exemplars of assessment categories. As participants examine student responses to the NAEP test items, the facilitator should listen carefully to the tone of the comments, taking care to remind the participants about issues of respect. Student work is shared for participants' own learning. Neither the facilitator nor participants should make assumptions about students or the conditions under which they learned.

Using the CD-ROM

To get an overview of the CD-ROM, see the CD contents list at the beginning of this manual.

This manual and the accompanying CD-ROM provide everything necessary to implement the workshops. Here you will find step-by-step instructions for workshop activities. On the CD-ROM you will find downloadable versions of the materials cited in the manual. You can also use the tools on the CD-ROM to modify the workshops by grade level and content. As you become more comfortable with the materials in the workshops, you may want to choose additional student responses or select different assessment items using the NAEP Item Search in the CD-ROM. Finally, you can construct your own workshop or add activities to the existing workshops by using the Activity Bank in the Resources section of the CD-ROM. Chapter 4 explains in detail how the CD-ROM supports the facilitation of the workshops in this manual.

Making the Connection

The *Learning From NAEP* materials work in concert with the IU-NAEP monographs and the materials made available by NCES on the NAEP Web site (http://nces.ed.gov/nationsreportcard/) to support both facilitators and participants as they continue their professional development beyond the workshop setting. These professional development materials are intended to empower educators to use standardized assessment data to better understand student learning and to improve teaching.

As you read through the manual, look for these icons to make connections among resources.

  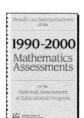

As educators become better acquainted with the tools and resources available to access and interpret assessment data, the manual will become less important and the CD-ROM, the NAEP Online Tools, and the monographs will become their primary resources. Ideally, educators will continue to use these dynamic resources to investigate the themes introduced in the workshops long after the sessions have ended.

2

REFERENCES

Kloosterman, P., & Lester, F. K., Jr. (Eds). (2004). *Results and interpretations of the 1990 through 2000 mathematics assessments of the National Assessment of Educational Progress.* Reston, VA: National Council of Teachers of Mathematics.

Lester, F. K., & Charles, R. I. (Eds). (2003). *Teaching mathematics through problem solving, prekindergarten–grade 6.* Reston, VA: National Council of Teachers of Mathematics.

Chapter 1
Introducing NAEP

Peter Kloosterman and Shelby P. Morge

O FTEN referred to as the "nation's report card," the National Assessment of Educational Progress (NAEP, usually pronounced "nape") was developed in the 1960s as a tool for monitoring precollege student performance in a variety of subject areas. The original design assessed students (ages 9, 13, and 17) as well as adults who were no longer in school (ages 17–35). The first mathematics assessment was completed in 1973, and subsequent assessments were administered in 1978, 1982, 1986, 1990, 1992, 1996, 2000, and 2003. In 1990, to make the testing process less burdensome for schools, NAEP moved from assessing students by age to assessing by grade level, and current results are reported on the basis of grade levels 4, 8, and 12. NAEP was designed so that each participating student is tested for about an hour and is given only a small subset of the items for his or her grade level. By pooling results from students, analysts can document progress for the nation as a whole.

> *NAEP is considered by many to be the most representative indicator of the mathematics skills that it tests, because of the large national sample size and the fact that the sample includes the lowest to the highest achieving students.*

 For a PowerPoint tutorial on the history and purpose of NAEP, as well as a discussion guide, visit the Activity Bank in the Resources section of the accompanying CD-ROM (Activity 1: Getting an Overview of NAEP).

NAEP does not provide information on individual student achievement, and thus NAEP is different from state and local accountability tests that are typically designed to identify students who need other kinds of work. Another difference between NAEP and state and local tests is the motivation level for students to complete the test and put in their best efforts. Because no individual scores are reported and students have no major incentives to do their very best work, we can reasonably assume that many students view NAEP as a low-stakes test. Thus, the suggestion has been made that NAEP may underestimate what students are able to do (O'Neil, Sugrue, & Baker 1996).

In spite of this shortcoming, in the United States NAEP is considered to be the most representative indicator of the mathematics skills that it tests, because of the large national sample size and the fact that the sample includes the lowest to the highest achieving students. NAEP has gained prominence in recent years because the No Child Left Behind (NCLB) legislation of 2001 specifies that NAEP may be used as a

No Child Left Behind (NCLB) Act of 2001— Legislation reauthorizing the Elementary and Secondary Education Act (ESEA), the main federal law affecting education from kindergarten through high school. NCLB is built on four principles: accountability for results, more choices for parents, greater local control and flexibility, and an emphasis on doing what works as verified by scientific research.

Main NAEP — An assessment instrument that reports information for the nation and specific geographic regions of the country, includes students drawn from both public and non-public schools, and reports results for student achievement at Grades 4, 8, and 12

State NAEP — A state level assessment that is identical in its content to Main NAEP but that selects separate representative samples of students for each participating jurisdiction or state because the national NAEP samples are not designed to support the reporting of accurate and representative state-level results

Long-Term Trend NAEP — Recurring assessment designed to give information on the changes in the basic achievement of United States youth; administered nationally, and reports student performance at ages 9, 13, and 17 in mathematics and reading; does not evolve on the basis of changes in curricula or in educational practices

The National Council of Teachers of Mathematics (NCTM) was the first professional organization to develop and release standards to raise expectations for student achievement in school mathematics (1989). Since the release of that document, test scores at Grades 4 and 8 on the National Assessment of Educational Progress reflect more than a grade level of improvement in performance. In 2000 NCTM issued *Principles and Standards for School Mathematics. Principles and Standards* is the compilation of the lessons learned and experiences gained over the 10 years of the 1990s and reflects the most current thinking, research, experience, and expertise of a wide variety of groups with an interest in mathematics education. Therefore, we draw on *Principles and Standards* for guidance throughout these materials.

benchmark to assess the extent to which state assessments are adequately determining student progress. In short, NAEP is quickly moving from being an assessment that has generated little interest among teachers and policymakers to one that could have a significant impact on school curricula, practice, and policy.

Structure of NAEP

Although the mathematics NAEP is often discussed as a single entity, it has evolved over time into three separate assessment programs: Main NAEP, State NAEP, and Long-Term Trend NAEP. Main NAEP documents student progress nationwide. Begun on a trial basis in eighth grade in 1990 and now an integral part of NAEP, State NAEP reports achievement for fourth and eighth grades on a state-by-state basis. The 2003 Main NAEP was a compilation of the results of all the State NAEPs, thereby permitting a comparison between individual states and nationwide achievement. Long-Term Trend NAEP has used the same items since the 1980s and allows for comparison of student performance today with that of 20 years ago. The main drawback of Long-Term Trend NAEP is that because it uses items that are 20 years old, it fails to adequately assess more recent additions to the curriculum, including statistics, algebraic thinking, and complex problem solving. All items in the *Learning From NAEP* professional development materials come from Main NAEP.

The mathematics content tested by NAEP and the types of items used are defined by the framework developed under the auspices of the National Assessment Governing Board (NAGB) (see Kenney, 2000; NAGB, 2002). The Main and State mathematics NAEPs include items from five content areas: (1) Number Sense, Properties, and Operations, (2) Measurement, (3) Geometry and Spatial Sense, (4) Data Analysis, Statistics, and Probability, and (5) Algebra and Functions. These five areas are similar to those in the NCTM *Curriculum and Evaluation Standards for School Mathematics* document (1989). However, the NAEP framework was constructed after much public input and is the result of consensus of the development committee rather than an intentional effort to align with the content of *Curriculum and Evaluation Standards*.

In addition to defining the five NAEP content strands, the 1996 and 2000 mathematics framework calls for the assessment of "mathematical power" at three ability levels (NAGB, 2002). According to NAGB, "mathematical power is characterized as a student's overall ability to gather and use mathematical knowledge through exploring, conjecturing, and reasoning logically; through solving nonroutine problems; through communicating about and through mathematics; through connecting mathematical ideas in one context with mathematical ideas in another context or with ideas from another discipline in the same or related contexts" (p. 35). To assess mathematical power, NAEP looks at students' abilities to reason in mathematical situations and to communicate in writing the logic used in solving problems. Some items also require students to connect mathematical ideas across contexts. The three interconnected ability levels specified by NAGB are "conceptual understanding," "procedural knowledge," and "problem solving" (Kenney, 2000; NAGB, 2002). The framework for the mathematics assessments conducted in 2005 and beyond introduces the concept of "level of complexity," which includes many of the elements of the ability levels of the previously used framework but focuses on item characteristics rather than on students' ability.

NAEP Test-Item Formats

For recent Main and State NAEP mathematics assessments, NAEP has used three item formats. The first is a standard multiple-choice format, with each 4th-grade item having four choices and each 8th- and 12th-grade item having five choices. NAEP does not use "all of the above" or "none of the above," so depending on grade level, four or five choices are offered, with a single correct response for each item. The second format, which has two variants, is called short constructed response (SCR). In the first variant, students simply write their answers in the space provided and are given full credit for a correct response and no credit for an incorrect response. In the second variant, students must answer multiple questions or provide a brief rationale for each response given. Students may earn partial credit on these items. The third type of item format is the extended constructed response (ECR). On these items, focused holistic rubrics allowing partial credit are used. ECR items are always placed at the end of the block of items to ensure that students do not spend so much time justifying answers that they do not get to the rest of the items. Overall, time is an issue, although not a major one, because at each grade level only a handful of items are left unanswered by a relatively large percentage of students. NAEP codes items that are skipped as "omitted" and identifies items that are not completed at the end of the assessment as "not reached."

NAEP Achievement Data

Since the 1986, NAEP, results have been reported as a scale score that is an overall indicator of student achievement at each grade level tested. The NAEP scaling system is independent of grade level, and scores can range from 0 to 500. This scaling system, based on item-response theory (see Braswell et al., 2001), allows scores on NAEP multiple-choice and constructed-response items of varying difficulty to be combined into a single, meaningful overall score. In addition to an overall score at each grade level, NAEP provides scale scores for each of the content strands (Measurement, Algebra and Functions, etc.) and by demographic characteristics. The scale scores for the content strands follow the same overall format as the composite scale scores. However, because of the way the scales in each area are defined, caution should be used in making comparisons across content areas. Thus, even though the score for Alge-

GLOSSARY

NAEP content strands — The five mathematics content areas used in the Main and State NAEP examinations: (a) Number Sense, Properties, and Operations; (b) Measurement; (c) Geometry and Spatial Sense; (d) Data Analysis, Statistics, and Probability; and (e) Algebra and Functions

NAEP ability levels — The general mental abilities associated with mathematics and targeted as primary foci in NAEP assessments: conceptual understanding, procedural knowledge, and problem solving

Short constructed-response item (SCR) — A type of short-answer item that requires students to (a) give either a numerical result or the correct name or classification for a group of mathematical objects, (b) draw an example of a given concept, or (c) write a brief explanation for a given result

Extended constructed-response item (ECR) — An item in which the situation demands more than a numerical or short verbal response, instead requiring the student to carefully consider a problem within or across the content strands, understand what is required to solve the problem, choose a plan of attack, carry out the attack, and interpret the solution in terms of the original problem

Multiple-choice item — An item that consists of one or more introductory sentences followed by a list of response options that include the correct answer and several incorrect alternatives

To see student responses to test items, visit the accompanying CD-ROM and click on the tab marked "NAEP Item Search." For example, you can find 30 examples of student responses to "Marcy's Dots" in the NAEP Item Search by entering the item number M054301, searching by grade level or content strand (eighth-grade Algebra and Functions), or pulling up the full list of test items for which student work is available and clicking on the item.

Marcy's Dots is just one of the released ECR items that Kenney found in the boxes containing NAEP booklets. In that test item, students are asked to explain how an imaginary student named Marcy might extend a pattern of dots to determine the number of dots in the 20th step without drawing and counting all the dots. In addition, students are asked to give the answer that Marcy should get for the number of dots. The item, from the Algebra and Functions strand, was administered to eighth-grade students in 1996. It was Item 9 in Block 1992-8M12; its individual number is M054301. Kenney copied numerous examples of student responses to Marcy's Dots. Those responses contained a variety of strategies.

bra and Functions may be higher than that for Geometry, the difference cannot appropriately be used as evidence that students are doing better in Algebra and Functions.

Student achievement on NAEP is also reported in terms of the number of students who meet the proficiency levels defined by the National Assessment Governing Board. The "basic" level indicates that students have acquired the minimal skills expected for their grade level, the "proficient" level indicates a solid understanding of the concepts and procedures for their grade level, and the "advanced" level indicates complete mastery of most if not all topics at their grade level. The No Child Left Behind legislation stipulates that all students must reach state-defined levels of proficiency by 2014, and thus those levels have become important from a policy prospective.

NAEP Student Responses

Most official NAEP reports of student performance are based on statistical, quantitative analyses of data (Silver, Alacaci, & Stylianou, 2000). Up to this point the richness of information about students' performance through a qualitative analysis of student responses has been largely untapped. However, the *Learning From NAEP* materials provide access to NAEP test items, results, *and* student responses so that participants can learn more about student understanding, teacher content knowledge, assessment, equity, and State NAEP issues. In particular, the materials support a series of workshops that address student performance on MC, SCR, and ECR items. The *Learning From NAEP* professional development materials were created with the hope that educators might use the five workshops as they exist in the manual and CD-ROM or use the suggestions and materials provided to build their own workshops for a specific grade level, content strand, or audience.

Although packets of student responses are often provided as part of these workshops, educators can find student responses for specific test items using a searchable database of student responses to NAEP SCR and ECR items—the NAEP Item Search. This tool, on the accompanying CD-ROM, contains almost a thousand student responses to items that have been given at the 4th-, 8th-, and 12th-grade levels over the NAEP assessment years. Student responses have been chosen that will best suit the needs of workshop facilitators and participants.

The *Learning From NAEP* collection of student responses emerged from the needs of a few mathematics educators. In the early 1990s, Patricia Kenney and her affiliates at the University of Pittsburg had access to NAEP scoring packets and the training packets given to raters. After looking through the packets, they decided they would like to see some student responses containing strategies beyond what had been addressed in the guides. Kenney made special arrangements with the National Center for Education Statistics to look at the secure student booklets, which were stored at the Pearson National Computing Services headquarters in Iowa City, Iowa. Upon arriving at the warehouse, she stood in awe as forklifts delivered palettes piled high with boxes of NAEP booklets. A tedious but worthwhile task lay ahead of her.

Kenney approached the selection of student responses with some strategies in mind to make the task less cumbersome. For example, she and her colleagues focused only on released items and consulted performance data to determine how many students actually responded to the item in a way that was scorable; items omitted by a high percentage of students were not considered. Kenney looked for NAEP booklets containing more than one SCR or ECR item. She spent an entire week using a copy machine in 1992 and returned to Iowa City for two more weeks in 1996. During this second visit, she was assisted by colleague Judith Zawojewski. Even though they had limited time to peruse each student response carefully, the goal was to find different strategies, misconceptions, and what Kenney described as "[potentially] just plain interesting work."

Without Kenney's determination to find "interesting" student responses, we would not have these valuable examples of student responses to use throughout our professional development materials. The NAEP student responses incorporated in these materials are helpful for mathematics educators in ways that are not addressed in the purpose of NAEP. For example, workshop participants may learn more about student understanding by looking at strategies, misconceptions, and different approaches to an item instead of focusing on overall NAEP results. In addition, workshop participants may feel more comfortable working with anonymous student responses than with the work of their own students. The student responses included in this CD-ROM were stripped of any identifiers. For this reason, workshop participants can easily develop judgment about the student work without critiquing or stereotyping the students. The student work included on the accompanying CD-ROM supplies workshop facilitators and participants with firsthand data to better understand student performance on NAEP and related assessments.

NAEP Professional Development

Although some of the features included in these workshops and accompanying CD-ROM can be found on the NAEP Web site and in other products, our inclusion of a large number of student responses for each item makes the *Learning From NAEP* professional development materials unique. The incorporation of student responses is significant because whereas the NAEP Web site provides only one or two exemplars of each NAEP scoring level, the *Learning From NAEP* CD-ROM has more than 30 student responses for each test item. Furthermore, these sets of student responses have been organized into workshops and activities that can be used in several different professional development situations. This collection of materials brings together the most recent

RESEARCH

"[Because of the scope of the project and because of limited time], I wasn't looking for frequency or quantity really, [but] different strategies, misconceptions, just interesting work." (Pat Kenney, personal communication, May 13, 2005)

GLOSSARY

NAEP Web site (online data tools) — World Wide Web site, www.nces.ed.gov/nationsreportcard/, maintained by the National Center for Educational Statistics, that has links to NAEP items, student performance, questionnaire results, and state-specific data

High-stakes test — An assessment instrument used to make significant educational decisions about students, teachers, schools, or school districts

data in mathematics achievement, a variety of tools to access those data, and background materials to help educators understand those data.

Learning From NAEP allows educators to learn more about NAEP and its current significance. Through these materials, educators gain an in-depth understanding of important issues in assessment, such as student understanding, content knowledge, and equity. The materials help teachers prepare students for high-stakes assessments. By investigating NAEP data, educators learn how performance on different assessments can reveal more about student learning. Chapter 2 focuses on the types of NAEP data that are available and how to interpret those data.

REFERENCES

Braswell, J. S., Lutkus, A. D., Grigg, W. S., Santapau, S. L., Tay-Lim, B., & Johnson, M. S. (2001). *The nation's report card: Mathematics 2000*. NCES 2001-517. Washington, DC: National Center for Education Statistics.

Kenney, P. A. (2000). "The seventh NAEP mathematics assessment: An overview." In E. A. Silver & P. A. Kenney (Eds.), *Results from the seventh mathematics assessment of the National Assessment of Educational Progress* (pp. 1–21). Reston, VA: National Council of Teachers of Mathematics.

National Assessment Governing Board. (2002). *Mathematics framework for the 2003 National Assessment of Educational Progress*. Washington, DC: National Assessment Governing Board.

National Council of Teachers of Mathematics (NCTM). (1989). *Curriculum and evaluation standards for school mathematics*. Reston, VA: NCTM.

National Council of Teachers of Mathematics (NCTM). (2000). *Principles and standards for school mathematics*. Reston, VA: NCTM.

O'Neil, H. F., Jr., Sugrue, B., & Baker, E. L. (1996). Effects of motivational interventions on NAEP mathematics performance. *Educational Assessment, 3,* 135–157.

Silver, E. A., Alacaci, C., & Stylianou, D. (2000). "Students' performance on extended constructed-response tasks." In E. A. Silver & P. A. Kenney (Eds.), *Results from the seventh mathematics assessment of the National Assessment of Educational Progress* (pp. 307–342). Reston, VA: National Council of Teachers of Mathematics.

Chapter 2
Understanding NAEP Data

Paula R. Stickles and Crystal Walcott

T HE RESULTS of the National Assessment of Educational Progress provide a wealth of data for United States school administrators and teachers. Regardless of numerous data dissemination outlets, many educators do not realize the valuable information that is easily retrieved through an Internet connection and a click of a mouse. By using the *Learning From NAEP* CD-ROM in conjunction with the NAEP online tools (www.nces.ed.gov/nationsreportcard/), teachers and administrators can investigate student performance statistics at both the item level and content-strand level as well as view a wide variety of student work samples of constructed-response items

Getting to Know NAEP Data

Once students participate in the NAEP assessment, the National Council of Education Statistics (NCES) begins work on compiling the data and preparing it for dissemination to the public. The United States public hears of general NAEP outcomes through national and state news outlets. For most people in the general population, the information provided in a brief news story suffices. For others interested in education policy, the detailed data available online provide an opportunity to examine achievement gains and gaps. Interested parties can perform their own mini-analyses by taking advantage of the display of comprehensive demographic data related to schools, teachers, and students. Appendix B of this manual presents a short primer on common data analysis procedures used with classroom and NAEP data, and works in concert with this chapter.

> *By using the* Learning From NAEP *CD-ROM in conjunction with the NAEP online tools, teachers and administrators can investigate student performance statistics at both the item level and content-strand level as well as examine a wide variety of student work samples of constructed-response items.*

Visit the NAEP Web site (http://nces.ed.gov/nationsreportcard/) to see the NAEP online tools referenced in this chapter: the NAEP Questions Tool, the NAEP Data Explorer, and the State Profiles Tool. This chapter focuses on interpreting the data obtained through the NAEP Questions Tool. For information about how to use the NAEP Data Explorer and State Profiles Tool, see chapter 3 of this manual.

GLOSSARY

NAEP Web site (online tools) — World Wide Web site, www.nces.ed.gov/nationsreportcard/, maintained by the National Center for Educational Statistics, that has links to NAEP items, student performance, questionnaire results, and state-specific data

Questions Tool — Online tool furnishing easy access to NAEP questions, student responses, and scoring guides that are released to the public; presents both national and state data, where appropriate

Data Explorer — An online tool offering tables of detailed results from NAEP national and state assessments using data based on information gathered from the students, teachers, and schools that participated in NAEP

State Profile — Online tool that presents important data about each state's student and school population and its NAEP testing history and results, providing easy access to all NAEP data for participating states and links to the most recent State Report Cards for all available subjects

11

Statistical Features of the NAEP Data Categories

As noted in the previous chapter, NAEP uses three types of questions, or test items. Short constructed-response (SCR) items are open-ended items requiring a brief written answer of a word or a phrase. NCES scores some NAEP SCR items as correct or incorrect. Extended constructed-response (ECR) items are also open-ended, but they require a more detailed written answer and are scored using a holistic rubric that sometimes allows partial credit. Multiple-choice (MC) items at the 4th-grade level contain four distractors, whereas 8th- and 12th-grade MC items include five distractors. At all grade levels, MC items are scored as correct or incorrect.

Occasionally, NAEP releases blocks of test items to the public. When an item is released, data related to the item are published on the NAEP Web site. One can search for a particular item by specifying the grade level, content strand, item type, and item difficulty. Along with the item itself, NAEP includes performance data, content-classification information, a scoring guide or key, and a few student responses for open-ended items. The user can also get more data disaggregated by state and by demographic subgroups. The item in Figure 2.1 was selected as a result of a search of the eighth-grade Algebra and Functions search using the NAEP Questions Tool.

Fig. 2.1. NAEP online tools: NAEP Questions Tool tabs.

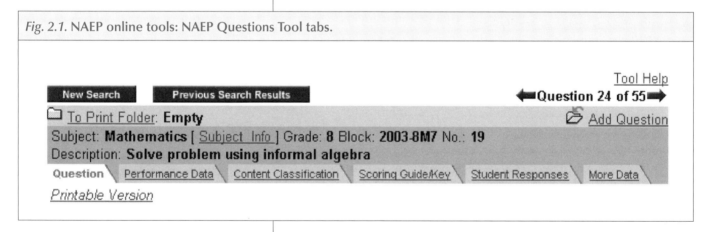

The data in figure 2.1 are also available on the *Learning From NAEP* CD-ROM in the NAEP Item Search, which contains a greater number and range of student responses. For example, the NAEP Web site provides only exemplars for each scoring category, whereas the NAEP Item Search on the CD-ROM supplies more than 30 student responses for each test item. In addition, the NAEP Item Search accesses printable versions for each of the NAEP Web site data categories: Performance Data, Content Classification, Scoring Guide with Student Exemplars, and More Data (see Figure 2.2).

To access the NAEP Questions Tool, one of the NAEP online tools, go to http://nces.ed.gov/nationsreportcard/itmrls/NQT_StartSearch.asp and search for any test item to see the data categories.

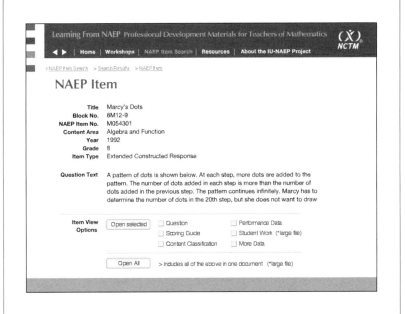

Figure 2.2. NAEP Item Search on the accompanying CD-ROM.

Analysts need to know how to use both tools. Although the NAEP Item Search on the accompanying CD-ROM offers easy access to data and additional samples of student responses, it may not contain the most recently released blocks of test items (the CD-ROM contains items released before January 2003). In addition, because of the focus on student responses, the test items on the CD-ROM are mostly of the ECR type. For the ECR and SCR items, the NAEP online Questions Tool provides only a few examples of student responses that correspond to the scoring categories in the rubric; however, it permits a search for all items that have been released. As a result, this chapter focuses on how to access and understand the data on the NAEP online tools while referencing the corresponding documents in the NAEP Item Search on the *Learning From NAEP* CD-ROM.

When you use the NAEP Questions Tool in the NAEP online tools (http://nces.ed.gov/nationsreportcard/itmrls/NQT_StartSearch.asp) or choose a test item from the Item Search on the accompanying CD-ROM, you find several headings or documents that relate to types of item-related data. The significance of these categories and the way the data are presented are important to understand. For example, the Performance Data tab reports general performance data in a bar graph (see Figure 2.3). The More Data tab offers specific data in table form (see Figure 2.4).

For a participant-friendly activity on how to access and use the NAEP online Questions Tool, visit the Activity Bank in the Resources section of the CD-ROM and choose Activity 2: "Using the NAEP Online Tools."

You can access the same data categories for test items on the accompanying CD-ROM by going to the NAEP Item Search and choosing an item. Each item is accompanied by a set of downloadable documents that match each of the NAEP Questions Tool categories.

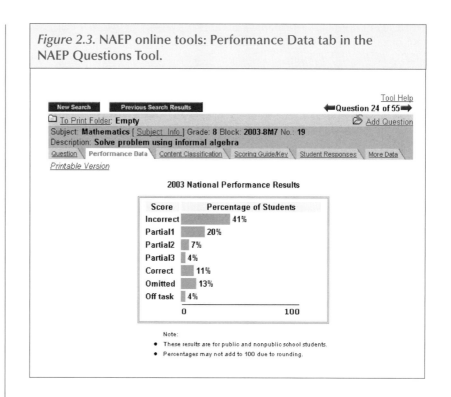

Figure 2.3. NAEP online tools: Performance Data tab in the NAEP Questions Tool.

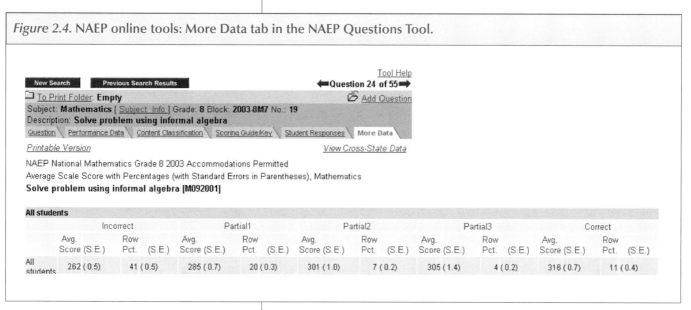

Figure 2.4. NAEP online tools: More Data tab in the NAEP Questions Tool.

Item-Level Data

NAEP ECR student responses are scored as extended, satisfactory, partial, minimal, or incorrect. NAEP provides additional data for the students who were off task or who omitted the item. Figure 2.4 is an example of an ECR item for which 11% of the students received an extended score while 4%, 7%, and 20% of the students received satisfactory, partial, and minimal scores, respectively. Additionally, 13% of the students omitted the item, and 4% of the student responses were off task. Performance data directly correspond with the Overall Performance table from the More Data tab.

Student responses to NAEP SCR and MC items are scored as "correct" or "incorrect," and some SCR items also can be scored as partially correct. The SCR Performance Data charts sometimes include information on the various types of incorrect responses that were obtained. The NAEP scoring guide defines the types of responses that fall into the different categories of incorrect responses. The More Data tab for SCR items and that for MC items contain data for the response categories shown in the Performance Data chart. The column with its heading marked with an asterisk (*) contains data for the correct response.

Item-Level Statistics

In the More Data category (Figure 2.4), the response category columns of the Overall Performance table contain three pieces of information: the average scale score, standard error (SE), and row percentage. To understand the data, one must understand that NAEP does not test all the nation's students in Grades 4, 8, and 12. Instead, NAEP uses a *representative sample* of students from across the nation. In 2003, the NAEP was administered to about 343,000 fourth- and eighth-grade students from nearly 13,000 U.S. schools (NCES 2005). Data compiled from the representative sample are used to make inferences about the entire population of students from which NAEP sampled. For instance, data from the sample of fourth-grade students is used to make inferences about the population of fourth-grade students as a whole; the same can be said for the eighth-grade sample.

Average scale scores

NAEP data are not reported in terms of raw scores. Instead, NCES converts raw NAEP scores into standardized scores. In this way, it establishes a common scale, allowing for a comparison of scores over time. The NAEP data available through the NAEP online Questions Tool provide the average scale score for students within the sample answering the given item with responses from each category.

No individual student scores are assigned to NAEP assessments; rather, NAEP scores reflect nationwide achievement. Of particular importance is the fact that no one student ever completes the entire NAEP assessment. Instead, students complete various blocks of items. Through a complex weighting process, NAEP statistically determines the expected score, ranging from 0 to 500, for a given student group. The scale score represents the average scale score of all the students who completed the entire set of questions. Keep in mind that the intention is not to assign individual student scores but to measure the nation's achievement as a whole.

Standard error

Since NAEP scale scores reflect the average score in a category from a sample of the nation's students, average scale scores vary somewhat by sample. In the instance of NAEP data, the sample's average scale score

GLOSSARY

Extended — The highest scoring level on an extended constructed-response item, indicating that the student's response matches the description given in the NAEP scoring rubric

Satisfactory — A scoring level on an extended constructed-response item, indicating that the student's explanation is adequate

Partial — A scoring level on an extended constructed-response item, indicating that the student's explanation is only to some extent correct

Minimal — A low scoring level on an extended constructed-response item, indicating that the student demonstrates very little understanding

Incorrect — The lowest scoring level on an extended constructed-response item, indicating that the student demonstrates no understanding

---

Scale score (Average scale score) — A numerical value, expressed on a scale of 0 to 500, derived from the overall level of performance of a group of students on individual NAEP assessment items; used in conjunction with interpretive aids, such as item maps, to provide information about what a particular aggregate of students in the population knows and can do

Standard error — A measure of sampling variability and measurement error for a statistic; estimated, because of NAEP's complex sample design, by jackknifing the samples from first-stage sample estimates; may also include a component that reflects the error of the measurement of individual scores estimated using plausible values

Row percentage — In a tabular presentation (such as in the NAEP Question Tool), the number of students represented in a particular cell of the table, divided by the number of students in the row of the table, converted to a percent

Representative sample — A portion of a population, or a subset from a set of units, that is selected by some probability mechanism for the purpose of investigating the properties of the population. NAEP does not assess an entire population, but rather selects a representative sample from the group to answer assessment items.

is used as a proxy for the population's achievement. The standard error indicates how much the sample mean is expected to vary from the population mean. As the standard error becomes larger, so does the expected variance of the sample mean from the population mean. The standard error of the mean is used in statistical tests of significance, such as a *t* test. Unfortunately, the NAEP Questions Tool does not offer this function, but a user can perform significance tests on NAEP data using one of the other NAEP online tools—the NAEP Data Explorer.

Row percentage

The final piece of information offered in the Overall Performance table is the row percentage, which is the percentage of students whose response to the item fell in the corresponding response category. For example, in Figure 2.4, 41% in the Incorrect column refers to the percentage of students who got the item incorrect.

Achievement-Level Data

Under the More Data tab, data are broken out into several categories: overall performance, achievement level, gender, parents' highest level of education, race/ethnicity as reported by the student (2002 and earlier), region of the country (2002 and earlier), race/ethnicity from school records (after 2002), schools public/nonpublic, schools public/Catholic/other nonpublic, and school location (1998 and earlier). These categories can reveal patterns and trends in achievement within and between categories. For example, let us take a closer look at achievement, ethnicity, and gender.

According to NCES (2004), achievement levels are intended to measure how students' actual achievement matches up against the desired achievement. The National Assessment Governing Board (NAGB) determines the achievement levels. The NAGB consists of teachers, education specialists, and members of the general public. By law, the achievement levels are in use on a trial basis and should be interpreted as such. However, NCES and NAGB believe that the levels are helpful in understanding the trends of student achievement.

The achievement levels are advanced, proficient, basic, and below basic. The advanced level is broadly defined as "superior performance." The proficient level represents "solid academic performance," and the basic level represents "partial mastery."

NAGB Levels	Scale Scores		
	4th	8th	12th
Basic	214	249	282
Proficient	262	299	333
Advanced	288	336	367

Consider the eighth-grade item-level data shown in Figure 2.5. Of the student group scoring at the advanced level (the group of students with scale scores of 336 and above), 87% answered this item correctly. If we refer to the Scoring Guide tab on the NAEP Web site or the corresponding document on the CD-ROM that accompanies this manual, we see that the student group falling within the advanced achievement level is made up of those students who are able to generalize and to develop models from examples and counterexamples. They are also able to use number and spatial sense to determine the reasonableness of an answer, to think abstractly, and to explain the reasoning behind their solution process.

Using the additional information from Figure 2.5, we see that 72% of the student group scoring at the proficient level (the group of students with scale scores between 299 and 336) answered the item correctly. The characteristics of students at the proficient level include the abilities to apply mathematical procedures consistently, to make conjectures, and to defend and explain their mathematical ideas. These students are also proficient in arithmetic and can use number and spatial sense in problem solving.

GLOSSARY

NAEP achievement levels — Performance standards—basic, proficient, and advanced—that measure what students should know and be able to do at each grade assessed by NAEP. The achievement levels are based on recommendations from panels of educators and members of the public, and provide a context for interpreting student performance on NAEP.

National Assessment Governing Board (NAGB) — An independent, bipartisan organization of individuals appointed by the U.S. Secretary of Education to give overall policy direction to the NAEP program. Its members include governors, state legislators, local and state school officials, educators, business representatives, and members of the general public.

Figure 2.5. CD-ROM NAEP Item Search: Achievement-level data in More Data.

ACHIEVEMENT LEVEL

	Incorrect #1				Incorrect #2				Correct				Omitted				Off task				Not Reached			
	Avg. Score	(S.E.)	Row Pct.	(S.E.)	Avg. Score	(S.E.)	Row Pct.	(S.E.)	Avg. Score	(S.E.)	Row Pct.	(S.E.)	Avg. Score	(S.E.)	Row Pct.	(S.E.)	Avg. Score	(S.E.)	Row Pct.	(S.E.)	Avg. Score	(S.E.)	Row Pct.	(S.E.)
Advanced	----	---	3%	0.8	343	2.2	9%	2.0	346	0.6	87%	2.3	----	---	----	---	----	---	0%	0.0	----	---	1%	0.5
Proficient	310	0.5	12%	0.8	312	0.6	15%	0.8	314	0.3	72%	0.9	310	1.6	1%	0.3	----	---	0%	***	311	1.1	2%	0.2
Basic	278	0.3	31%	0.7	280	0.5	19%	0.7	283	0.3	47%	0.8	278	1.4	3%	0.3	----	---	0%	0.1	278	0.9	4%	0.3
Below Basic	234	0.6	56%	0.9	241	1.0	17%	0.6	248	0.6	16%	0.7	229	1.2	10%	0.6	220	4.6	1%	0.1	231	2.8	7%	0.4

The remaining data in Figure 2.5 give us additional achievement-level information. Of the group of students scoring at the basic level (scale scores of 249 to 299), 47% answered the item correctly. The mathematical characteristics of this student group include the abilities to demonstrate procedural and conceptual understanding of arithmetic and to solve problems with structural prompts, such as diagrams and charts. Students in this group show a limited ability in mathematical communication (NCES, 2002). The data from Figure 2.5 indicate that 16% of the student group scoring at the below-basic level (scale scores below 249) answered this item correctly.

Data are also reported on the basis of ethnicity. Race is self-reported by students in assessments up until 2002 and reported by the school after 2002. The data in Figure 2.6 reveal the percentage of students in each race answering correctly, incorrectly, omitting the problem, and so on. For instance, for this item, 31% of American Indians answered correctly, and 7% of Hispanic students did not reach the item.

On the NAEP Web Site, the Data Explorer More Data tab furnishes specific national and state NAEP data but no longer gives the achievement level data (see Figure 2.5). You can still view achievement level data for selected items by clicking on More Data in the NAEP Item Search on the CD-ROM.

Activity 3, "Understanding Data," from the Activity Bank in the Resources section of the CD-ROM gives participants opportunities to work tasks that examine central tendency and significance within NAEP and classroom data.

Note that the data presented in Figures 2.5 and 2.6 contain the categories Incorrect #1 and Incorrect #2. The difference between these two categories is that Incorrect #1 represents a particular common error that arises in some responses to the question. NAEP tracks this common error separately from the other incorrect solutions. Incorrect #2 corresponds to all other incorrect solutions. To determine the total percentage of incorrect responses, add the values found in the Incorrect #1 and Incorrect #2 columns. Occasionally, a cell may contain no data, meaning that sufficient data do not exist to make estimates for the particular response category.

Figure 2.6. NAEP online tools: Race/ethnicity data in More Data in the NAEP Questions Tool.

RACE/ETHNICITY used in 2002 & 2003 NAEP reports

	Incorrect #1		Incorrect #2		Correct		Omitted		Off task		Not Reached	
	Avg. Score (S.E.)	Row Pct. (S.E.)	Avg. Score (S.E.)	Row Pct. (S.E.)	Avg. Score (S.E.)	Row Pct. (S.E.)	Avg. Score (S.E.)	Row Pct. (S.E.)	Avg. Score (S.E.)	Row Pct. (S.E.)	Avg. Score (S.E.)	Row Pct. (S.E.)
White	267 0.7	27% 0.5	283 0.9	17% 0.4	301 0.4	54% 0.6	257 2.7	2% 0.2	---- ---	0% 0.0	276 2.1	3% 0.2
Black	243 1.5	47% 1.1	256 1.4	18% 0.8	278 1.1	25% 1.2	238 2.5	9% 0.6	---- ---	0% 0.1	247 2.5	10% 0.7
Hispanic	248 0.9	46% 1.7	265 2.3	15% 1.1	282 1.6	28% 1.4	243 3.4	10% 0.8	---- ---	1% 0.1	248 5.0	7% 0.6
Asian/Pacific Islander	268 3.7	31% 2.2	293 4.1	16% 1.8	310 2.2	47% 2.5	258 6.4!	6% 1.1	---- ---	0% 0.1	262 5.5	2% 0.5
American Indian	251 2.7	43% 3.6	262 6.0	19% 2.6	289 4.5	31% 2.7	---- ---	7% 2.3	---- ---	1% 0.5	---- ---	5% 1.2

RESEARCH

Whenever feasible, "NAEP is to collect, cross tabulate, compare and report information by disability and limited English proficiency in addition to race, ethnicity, socioeconomic status and gender." No Child Left Behind Act of 2001, P.L. 107-110, 115 Stat. 1425 (2002).

Variations in Data Format

Let us revisit the Marcy's Dots item from chapter 1. Since the item is from 1992, the data do not exactly match the form of the foregoing data tables. Since the Web site has been restructured over the years, items administered before 2003 vary somewhat in the data tables. All the data reported are the same, but they appear in a slightly different format. Consider the data for Marcy's Dots in Figures 2.7 and 2.8.

The performance data for the item are in Figure 2.7. Notice that the data in Figure 2.7 do not appear to correspond to the data in Figure 2.8. This example pinpoints some of the variations that exist in the NAEP Web site. The categories appear in reverse order under the More Data tab. The extended category corresponds to column D, the satisfactory category corresponds to column C, the partial category corresponds to column B, the minimal category corresponds to column A, and the incorrect/off task and omitted-item categories are combined and correspond to the Omitted column. Note that the column titled Not Reached does not appear in the Performance Data tab.

Variations such as described occur throughout the older data items. However, the correspondence between the Performance Data tab and the More Data tab is straightforward and can be determined with little difficulty. Since the Performance Data and the More Data documents on the CD-ROM were drawn from the NAEP Web site, they also reflect those variations in format.

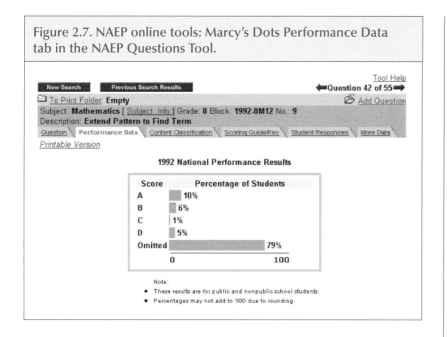

Figure 2.7. NAEP online tools: Marcy's Dots Performance Data tab in the NAEP Questions Tool.

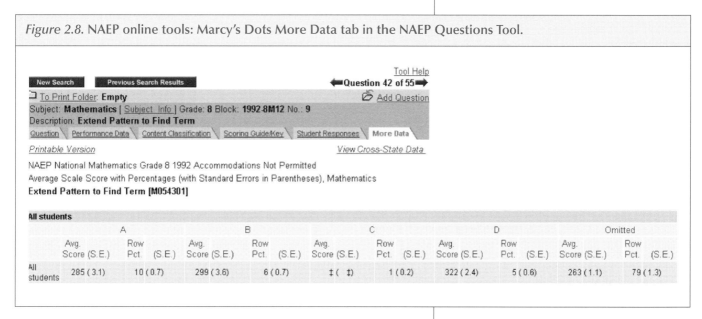

Figure 2.8. NAEP online tools: Marcy's Dots More Data tab in the NAEP Questions Tool.

Take the Next Step

Although this chapter helps you make sense of the NAEP item level and scale score data available from the NAEP Questions Tool, the next chapter will lead you step-by-step through the process of accessing the data electronically from all three of the NAEP online tools: the NAEP Questions Tool, the NAEP Data Explorer, and the State Profiles Tool.

By using the data available on the NAEP Web site and the *Learning From NAEP* CD-ROM, teachers and administrators can compare the achievement of their classes and districts with that of the nation as a whole. Remember that although the NAEP online Questions Tool allows you to search for items released in the years 1990, 1992, 1996, and 2003, the online samples of student work for constructed-response items are limited to an exemplar for each response category. In contrast,

In addition to mathematics data, the NAEP Web site (www.nces.ed.gov/nationsreportcard/) provides information related to the nation's achievement in reading, writing, science, U.S. history, civics, geography, and the arts. In each discipline you will find data related to student performance and released assessment items. Also, the Web site features results from the NAEP Long-Term Trend assessments. Unlike Main NAEP, which changes over time to reflect current instructional practices, the Long-term Trend (LTT) NAEP remains stable over time, allowing for the analysis of trends in results over several years. NAEP recently released a number of LTT items to the public, making them available electronically on the NAEP Web site.

the CD-ROM that accompanies this volume contains a wealth of student responses across all categories corresponding to items released prior to 2003. By combining the information on the CD-ROM with that from the NAEP online tools, teachers and administrators can investigate student performance statistics at both the item level and content-strand level while examining a wide variety of mathematical strategies as they are represented in student responses to constructed-response items.

REFERENCE

National Center for Education Statistics. (2002, 2004, 2005). *The NAEP-mathematics achievement levels by grade.* Retrieved from http://nces.ed.gov/nationsreportcard/mathematics.

Chapter 3
Exploring the NAEP Online Tools

Peter Kloosterman, Myoungwhon Jung, and YoungOk Kim

THE NAEP data have been available to the public since 2001, when Educational Testing Service (ETS) and NCES staff provided easy access to it through the NAEP Web site (http://nces.ed.gov/nationsreportcard/). The Web site furnishes three tools for obtaining NAEP data. In the previous chapter, we presented an overview of the data provided by the NAEP Questions Tool and how it relates to the materials on the *Learning From NAEP* CD-ROM. This chapter focuses on the details of using the NAEP Web site online tools, specifically the NAEP Questions Tool, the NAEP Data Explorer, and the NAEP State Profiles Tool. These tools are found on the NAEP Web site under the headings Sample Questions, Analyze Data, and State Profiles (see Figure 3.1).

According to the NAEP Web site, the NAEP online tools are designed to work with Internet Explorer and Netscape browsers, although Internet Explorer is recommended. From our experience, both work well on a PC platform, but on a Macintosh platform, Netscape is the only browser that makes the NAEP Data Explorer fully functional.

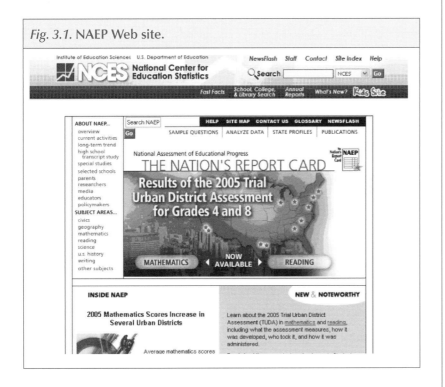

Fig. 3.1. NAEP Web site.

Overview of Web-Based Data Tools

The Questions Tool allows anyone to view items released by NCES. It presents performance data, scoring guides, and student work exemplars for released test items. The Data Explorer allows the user to view NAEP

Activity 2, "Using the NAEP Online Tools," from the Activity Bank in the Resources section of the CD-ROM provides a step-by-step tutorial to help participants learn how to use the three NAEP online tools.

 The Advanced Search option is available on the CD-ROM's NAEP Item Search. As noted previously, the primary difference between the NAEP online Questions Tool and the NAEP Item Search on the CD-ROM is the number and type of student responses available on the CD-ROM. Rather than the 5 or 6 exemplars available on the NAEP Web site, the NAEP Item Search on the CD-ROM furnishes 20 to 30 student responses that are not scored, thus offering examples of mathematical strategies not necessarily evident from the online exemplars or scoring rubrics.

GLOSSARY

Main NAEP — An assessment instrument that reports information for the nation and specific geographic regions of the country, includes students drawn from both public and non-public schools, and reports results for student achievement at Grades 4, 8, and 12

Long-Term Trend NAEP — Recurring assessment designed to give information on the changes in the basic achievement of United States youth; administered nationally, and reports student performance at ages 9, 13, and 17 in mathematics and reading; does not evolve on the basis of changes in curricula or in educational practices

Quick Search — A search for NAEP questions by subject and grade, available as a data tool on the CD-ROM in the NAEP Item Search section

Advanced Search — A search for questions by grade, content classification, question type, difficulty, and other subject-specific variables

results for the nation as a whole, by state or region, or in relation to numerous background questions given to participating students, teachers, and schools. State Profiles are available for individuals who want to investigate a state's student and school population and its NAEP testing history and results. All three NAEP online tools have corresponding materials on the accompanying CD-ROM and are highlighted in the margin when appropriate.

NAEP Questions Tool

Begin by going to the NAEP Web site (http://nces.ed.gov/nationsreportcard/) and clicking on the heading Sample Questions (see Figure 3.1). This selection takes one to the NAEP Questions page. Click on Search Options, and choose Quick Search, Advanced Search, or Search by Block (Main NAEP only) under the headings Main NAEP Questions and Long-Term Trend Questions.

For the Main NAEP questions, Quick Search is the easiest way to find NAEP test items, unless items from specific mathematics content areas are desired. Selecting Mathematics and a grade level and then clicking Search brings up a list of items that link to the individual test-item data.

Advanced Search allows the user to narrow the search by content areas (Number Sense, Properties, and Operations; Measurement; Geometry; Data Analysis and Probability; Algebra), question types (multiple choice, short constructed response, extended constructed response), mathematical complexity (low complexity, moderate complexity, high complexity), mathematical ability (conceptual understanding, procedural knowledge, problem solving), and question difficulty (easy, medium, hard). The user does not have to make a selection in every category to complete a search (see Figure 3.2).

Searching by block finds specific blocks of released items sorted by subject area and grade. Blocks of items appear together in a particular NAEP test booklet.

All three search options yield a list of questions in a table. For example, going to Advanced Search Options; choosing 8th Grade, Algebra, and Extended Constructed Response; then clicking on Search brings up a page that resembles Figure 3.3.

The Questions Tool item page lists the items that match search constraints in a table; in Figure 3.3 the search returned three items. In addition to the subject area and grade level for each item, the table includes a short description of the item, the year and the block when the item was last used, and the item number in that block. Clicking on the description for each item results in a page that shows the actual test item along with five tabs that can be used for retrieving more data about the item. For example, clicking on the item description Extend Pattern to Find Term (Marcy's Dots problem) in the table results in a display of the actual item, but more data about the item can be obtained by clicking

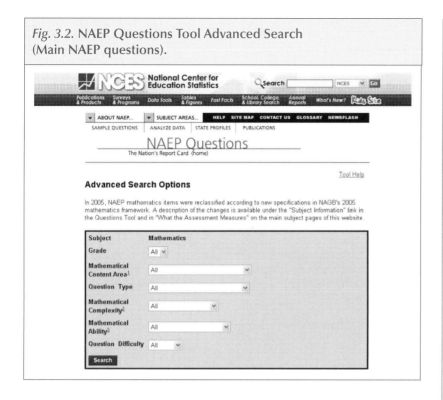

Fig. 3.2. NAEP Questions Tool Advanced Search (Main NAEP questions).

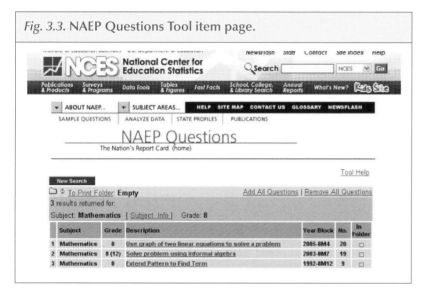

Fig. 3.3. NAEP Questions Tool item page.

the tabs labeled Performance Data, Content Classification, Scoring Guide/Key, Student Responses, and More Data (Figure 3.4). For more information about how to use the data categories on the item page, review chapter 2.

Most of the information displayed by the various tabs can be formatted for effective printing, and groups of individual items can be collected in a To Print Folder. First, click on the Add Question option, which is located at the right side of the item-information page. Then click on the To Print Folder to choose information to display or print out. Next select the information to be printed, and press Assemble Document. When choosing more than one item, select Edit Options (Figure 3.5) to specify the order in which items appear.

GLOSSARY

Search by block — Within a subject, a feature that allows searching for a specific block, or booklet, of questions that were administered to students who participated in the NAEP assessment

Block — A group of assessment items created by dividing the item pool for an age or grade into subsets; used in the implementation of the BIB spiral sample design

Item — The basic, scorable part of an assessment; a test question

Performance data — Information that reveals the percentage of students at each score level

Content classification — A designation of the mathematical content area(s) and knowledge that the selected item assesses

Scoring guide/key — A list of the correct answers for multiple-choice questions; also, a holistic rubric used to score each student's response for short constructed-response and extended constructed-response items

Student responses — Student answers that have been scored on the basis of the NAEP scoring rubric; not available for multiple-choice questions

More data — Additional data for each NAEP item, including the performance of various subgroups on the item

The NAEP Item Search on the CD-ROM also permits a search by item number, grade level, and content area.

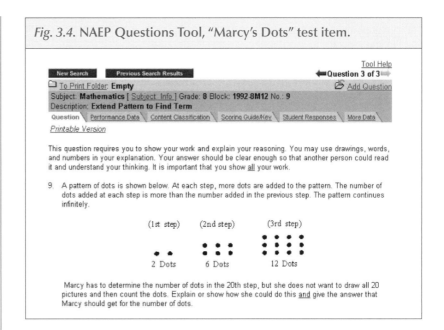

Fig. 3.4. NAEP Questions Tool, "Marcy's Dots" test item.

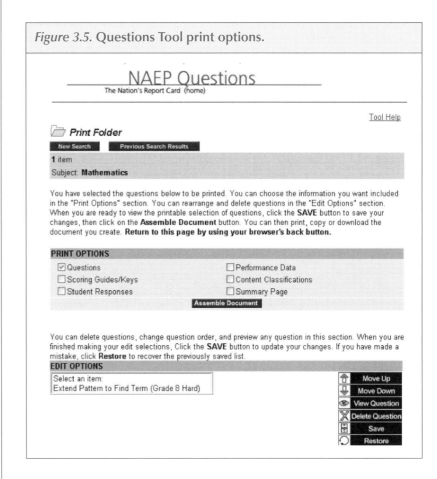

Figure 3.5. Questions Tool print options.

One can also print multiple questions from the table (see Figure 3.3). To begin, add questions to the To Print Folder by checking boxes under In Folder. After selecting all desired questions, click on To Print Folder to bring up the next window. Check the information to be printed, then click Assemble Document. Again, to decide the order in which items appear, select Edit Options (Figure 3.5).

For those interested in viewing the Long-Term Trend NAEP questions, Quick Search is the easiest option. Because relatively few Long-Term Trend items are available to the public, choosing Mathematics and an age results in a small set of items that can easily be reviewed.

Advanced Search allows the user to narrow a search of Long-Term Trend items by choosing an age (9, 13, 17), a content area (number and numeration; variables and relationships; shape, size, and position; measurement; probability and statistics), and a question type (multiple choice, short constructed response, extended constructed response) (see Figure 3.6). Clicking on the description for each item in the list brings up the page that presents the actual test item along with scoring and performance information similar to that available for Main NAEP items.

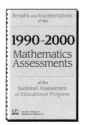

To find more information regarding the Long-Term Trend mathematics assessment, see chapter 7 of Kloosterman & Lester (2004, pp. 175–191). A summary of chapter 7 is also found on the accompanying CD-ROM in the About the IU-NAEP Project section.

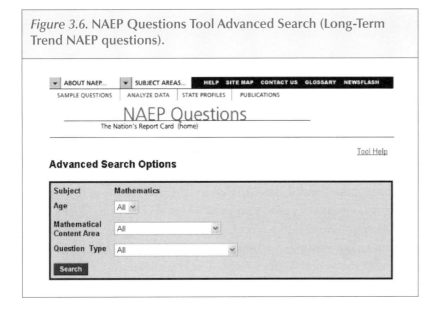

Figure 3.6. NAEP Questions Tool Advanced Search (Long-Term Trend NAEP questions).

NAEP Data Explorer

One can access the NAEP Data Explorer by going to the NAEP Web site (http://nces.ed.gov/nationsreportcard/). Clicking on Analyze Data yields Quick Start and Advanced options. Quick Start mode is useful for convenient access to the major reporting variables used in NAEP reports. The user needs to specify five parameters: grade, subject, jurisdiction, variable, and year (see Figure 3.7). The options for jurisdiction are national, national (public), or any state or territory. The national option includes students from public and nonpublic schools, whereas state-level results, because of limited nonpublic school samples in some states, are given only for public schools. The national (public) option is appropriate for looking at national data in relation to state-level data.

GLOSSARY

Public and nonpublic schools — The types of school that students attend, and according to which the results of NAEP assessments are reported. Nonpublic schools include Catholic and other private schools. Because they are funded by federal authorities, not state/local governments, Bureau of Indian Affairs (BIA) schools, Department of Defense Domestic Dependent Elementary and Secondary Schools (DDESS), and Department of Defense Dependents Schools (Overseas) are not included in either the public or nonpublic categories; they are included in the overall national results.

Figure 3.7. NAEP Data Explorer Quick Start.

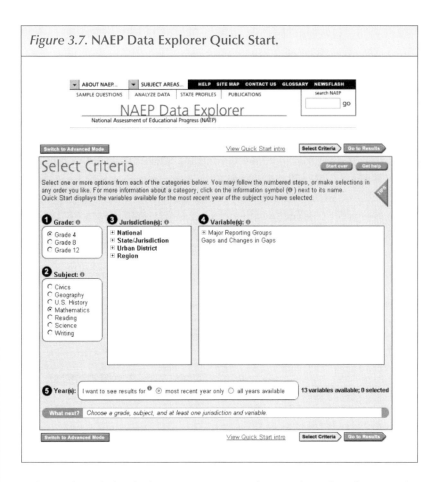

Advanced mode leads the user to a comprehensive list of student, teacher, and school variables and, compared with Quick Start mode, includes more options for years to be considered and more formats for printing out results. For example, to determine whether fourth-grade students agree with the statement "Learning mathematics is mostly memorizing facts," go to the Data Explorer's Advanced mode and select Grade 4, Mathematics, National (national is the most inclusive measure, although other options can be used if desired), and Student Factors (students' perceptions of mathematics are within the Student Factors category) (see Figure 3.8). Under the Student Factors category, clicking on Affective Disposition generates a list of student factors including the factor "Math is mostly memorization." The factor is listed twice, once followed by <2003 and once followed by 2003 (Figure 3.9). The 2003 data on this item are listed separately because, although the wording of the question was the same in 2003 as it was in prior years, the response options for 2003 were "agree," "not sure," and "disagree" whereas the options for prior years were "agree," "undecided," and "disagree." Because of this difference, one must be careful comparing results for 2003 with those of previous years.

Learning From NAEP

Figure 3.8. NAEP Data Explorer Advanced mode.

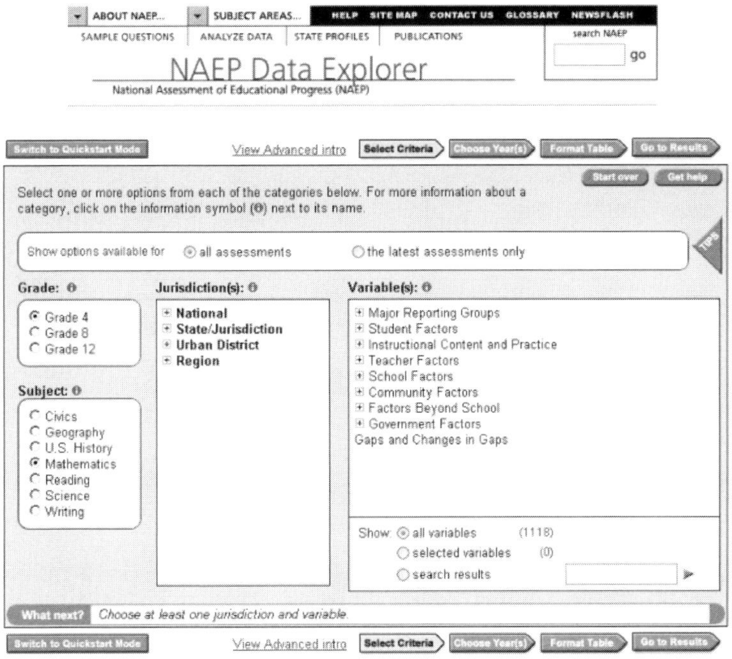

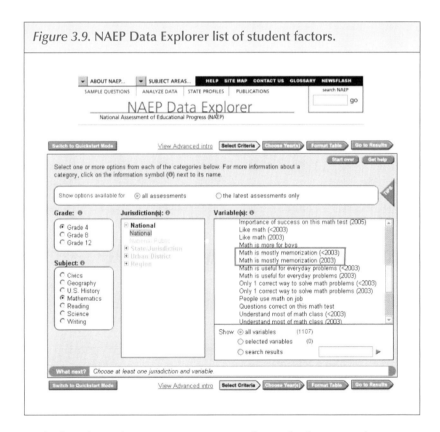

Figure 3.9. NAEP Data Explorer list of student factors.

To look at data relating to memorization for multiple years, select "Math is mostly memorization (<2003)" and click on one of the

GLOSSARY

Accommodations — Alterations in the administration of standardized assessments, such as NAEP, that are provided to certain students with disabilities (SD) or limited English proficiency (LEP), as specified in the student's Individualized Education Program (IEP)

Statistically significant — A term used to indicate that the observed differences are likely due to true differences between groups rather than sampling and measurement error

Choose Year(s) arrows (Figure 3.10) to generate a table in which to select desired years. A superscript 1 follows some of the years to signify "accommodations not permitted," meaning that the data do not include any students who received special treatment during the assessment (e.g., being allotted more time or having items read to them) because of their special needs.

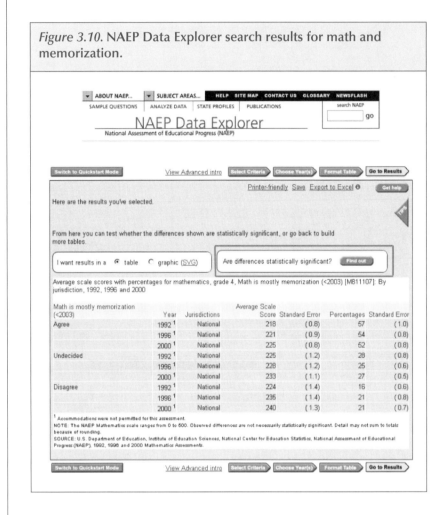

Figure 3.10. NAEP Data Explorer search results for math and memorization.

Assume one wanted to know more about scale scores and percentages of students for who no accommodations were permitted in relation to mathematics and memorization for years prior to 2003. On the table showing years available, select the boxes for 2000[1], 1996[1], and 1992[1]. Click on Format Table, select average scale score with percentages, and then click on one of the Go to Results arrows to bring up the table displayed in Figure 3.10. As can be seen in that table, 52% of fourth graders agreed that mathematics is mostly memorizing, 27% were undecided, and 21% disagreed. The average scale score of Year 2000 students who agreed was 225, the average scale score of those who were undecided was 233, and the average scale score of those who disagreed was 240. In other words, the more strongly students disagreed that learning mathematics is mostly memorizing facts, the higher they scored. The table also shows that the percentage of students who agreed with the statement decreased from 1992 to 1996 (57% to 54%) and again in 2000 (52%). If one wanted to know, for example, whether the same pattern was true with respect to geometry rather than mathematics in general, going back to the Format Table screen and selecting geometry in the Subscale window leads to the desired information.

One final important option on the Format Table screen is called Show Long Titles. Clicking on that box gives the exact statement that students responded to rather than a shortened version. On items such as the one we have been discussing, the exact statement ("Learning math is mostly memorizing facts") has a somewhat different meaning than the shortened form used on most screens ("Math is mostly memorization").

Whenever statistics are used to compare groups of students, the possibility arises that a higher score for one group is the result of random chance rather than representative of true differences between groups. On the Data Explorer search results screen, the question "Are differences statistically significant?" appears above the results table (Figure 3.10). Clicking on Find Out brings a pop-up window that lets the user check differences in average scale scores or the percentage of students giving each response within a year or across years. To determine whether the relation between agreement and scale score is statistically significant, choose 2000 from the Year options and Average Scale Score from the Statistic Type options, and then click on Compute.

The top row of the table in the pop-up window (Figure 3.11) shows that the average scale score of students in agreement is significantly lower than the average scale score of students who are undecided or in dis-

Activity 3, "Understanding Data," from the Activity Bank in the Resources section of the CD-ROM presents a series of tasks that examine central tendency and significance within NAEP classroom data

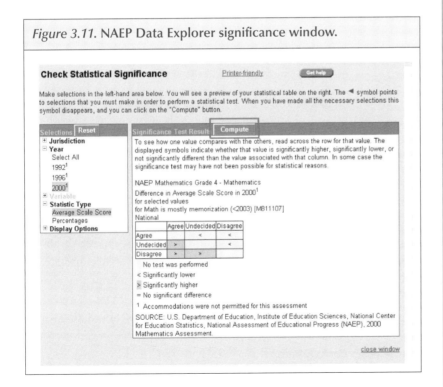

Figure 3.11. NAEP Data Explorer significance window.

The "Are differences statistically significant?" option on the results window (Figure 3.10) allows the user to check for significant differences across jurisdictions, years, and variables. Selecting Show Detail under Display Options (Figure 3.11) allows users to see the *p*-values (probability of true differences) in each comparison.

GLOSSARY

NAEP achievement levels — Performance standards—basic, proficient, and advanced—that measure what students should know and be able to do at each grade assessed by NAEP. The achievement levels are based on recommendations from panels of educators and members of the public, and provide a context for interpreting student performance on NAEP.

agreement. The middle row shows that the average scale score of those who are undecided is significantly lower than the average scale score of those who are in disagreement.

Reading the pop-up window in relation to the numbers in the main window in which average scale scores are displayed helps to clarify the differences shown in the pop-up window. For example, the main window (Figure 3.12) shows that the Year 2000 scale score (225) for students who agreed with the statement was lower than the scale scores of their counterparts who were undecided (233) or who disagreed (240). The less-than sign in the table in the pop-up window shows that the difference is statistically significant. Similarly, those who were undecided or who disagreed scored significantly higher than those who agreed.

Figure 3.12. NAEP Data Explorer significance (pop-up and main window).

Math is mostly memorization (<2003)	Year	Jurisdictions	Average Scale Score	Standard Error	Percentages	Standard Error
Agree	1992 [1]	National	218	(0.8)	57	(1.0)
	1996 [1]	National	221	(0.9)	54	(0.8)
	2000 [1]	National	225	(0.8)	52	(0.8)
Undecided	1992 [1]	National	225	(1.2)	28	(0.8)
	1996 [1]	National	228	(1.2)	25	(0.6)
	2000 [1]	National	233	(1.1)	27	(0.5)
Disagree	1992 [1]	National	224	(1.4)	16	(0.6)
	1996 [1]	National	235	(1.4)	21	(0.8)
	2000 [1]	National	240	(1.3)	21	(0.7)

National

	Agree	Undecided	Disagree
Agree		<	<
Undecided	>		<
Disagree	>	>	

No test was performed

< Significantly lower

> Significantly higher

= No significant difference

[1] Accommodations were not permitted for this assessment

To test whether significantly fewer students each year are agreeing that mathematics is memorizing facts, close the pop-up window and go back to "Are differences statistically significant?" in the results window. Choosing Select All in the Year option, Agree in the Variable option, and Percentages in the Statistic Type option results in a table that shows that the percentage of students who agreed that mathematics is mostly memorizing facts did not decrease significantly from 1996 to 2000 (54% as compared with 52%) but that a statistically significant drop in percentage occurred among those who agreed in 1992 (57%) as compared with 2000 (52%).

NAEP State Profiles Tool

Access the NAEP State Profiles Tool by going to the NAEP Web site (http://nces.ed.gov/nationsreportcard/) and clicking on State Profiles. An interactive map (Figure 3.13) appears, and the user can choose a state either by clicking on the state on the map or by selecting the state name from the window above the map.

To obtain a detailed State Report Card in PDF format, click on Mathematics: Grade 4 or Grade 8, which is located below the table titled "History of NAEP Participation and Performance" on the chosen State Profile page.

The "Exploring State NAEP" workshop (chapter 8 in the manual and on the accompanying CD-ROM) uses the State Profile Tool to examine how the mathematics score gaps have narrowed or widened in targeted states across assessment years.

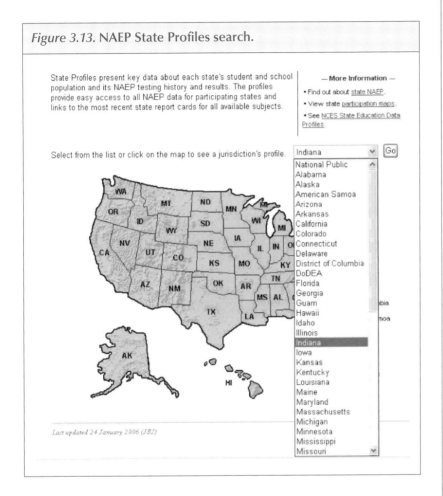

Figure 3.13. NAEP State Profiles search.

The State Profiles tool provides access to NAEP data about each state's student and school population, including links to the most recent state report cards. Note that state data are available only for Grades 4 and 8—no State NAEP is administered at Grade 12. Following a state's link leads to a wide range of demographic and achievement information for all 50 states plus U.S. territories. Indiana data, for example, are shown in Figure 3.14. Results can also be viewed in a graphic format using the options on the right side of the table (see Figure 3.14). The options also include cross-state maps and tables showing how a focal state compares with the other states in the country. Clicking on the links under the option of Cross-State Comparison Maps generates two graphic and table-format options. Clicking on one of the check marks in the chart that appears leads to a comparison graphic or table (see Figure 3. 15). At the bottom of each State Profile page is a link that takes the user to the NAEP Data Explorer for more detailed state results.

The task of examining differences between groups is sometimes easier if one views the state data as a graphic. The NAEP Data Explorer (www.nces.ed.gov/nationsreportcard) will create graphic data displays if the user has SVG (Scalable Vector Graphics) viewer software installed. This software can be downloaded at no cost from www.adobe.com/svg.

State Report Cards, or State Snap Shots, are also available in PDF format in the Resources section of the accompanying CD-ROM. State Gaps Reports that provide summary data for selected subgroups are also available for download in the same section.

Figure 3.14. One part of the results of a NAEP State Profiles search for Indiana.

History of NAEP Participation and Performance

Subject	Grade	Year	State Avg.	[Nat. Avg.]	Basic	Proficient	Advanced	Graphics
Mathematics (scale: 0-500)	4	1992ʰ	221	[219]	60	16	1	• Scale Scores
		1996ʰ	229	[222]	72	24	2	• Achievement Levels
		2000	233	[224]	77	30	2	• Cross-State Comparison Maps:
		2003	238	[234]	82	35	4	○ Scale Scores
		2005	240	[237]	84	38	5	○ Percent at or Above Proficient
	8	1990ʰ	267	[262]	56	17	3	
		1992ʰ	270	[267]	60	20	3	
		1996ʰ	276	[271]	68	24	3	
		2000	281	[272]	74	29	5	
		2003	281	[276]	74	31	5	
		2005	282	[278]	74	30	5	
Reading (scale: 0-500)	4	1992ʰ	221	[215]	68	30	6	• Scale Scores
		1994ʰ	220	[212]	66	33	7	• Achievement Levels
		2002	222	[217]	68	33	7	• Cross-State Comparison Maps:
		2003	220	[216]	66	33	8	○ Scale Scores
		2005	218	[217]	64	30	7	○ Percent at or Above Proficient
	8	2002	265	[263]	77	32	2	
		2003	265	[261]	77	33	3	
		2005	261	[260]	73	28	2	
Science (scale: 0-300)	4	2000ʰ	155	[148]	75	32	3	• Scale Scores
								• Achievement Levels
	8	1996ʰ	153	[148]	65	30	2	• Cross-State Comparison Maps:
								○ Scale Scores
		2000ʰ	156	[149]	68	35	3	○ Percent at or Above Proficient
Writing (scale: 0-300)	4	2002	154	[153]	88	26	1	• Scale Scores
								• Achievement Levels
								• Cross-State Comparison Maps:
								○ Scale Scores
	8	2002	150	[152]	85	26	1	○ Percent at or Above

Figure 3.15. NAEP State Profiles: Cross-state comparison for Indiana.

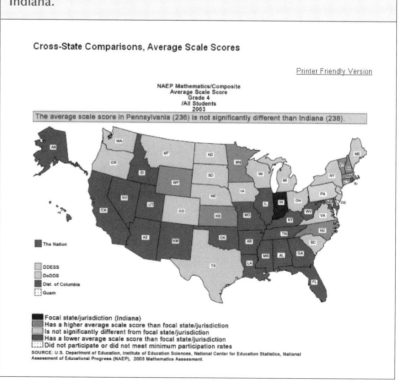

Nathan Walton and Jeff Hanson

THE COMPANION *Learning From NAEP* CD-ROM is located at the back of this manual. The goals of the CD are threefold: (1) to furnish facilitators and workshop participants with the materials to complete the activities in the workshops (chapters 5–10); (2) to give facilitators and participants the tools to create their own workshops; and (3) to provide access to NAEP data, student responses, and NAEP-related resources for personal professional development.

CD-ROM Overview

The accompanying CD is compatible with both Microsoft Windows and Mac OS X. Simply insert the CD into the computer drive, and follow these directions:

- For Windows: The CD is programmed to begin automatically. If it does not, simply go to the root directory on the CD and click the IU-NAEP.exe application (CD/IU-NAEP.exe).

- For Macintosh: Look for the LEARNING FROM NAEP CD on the desktop. Open the CD, and click on IU-NAEP.app.

The first screen that appears after opening the CD-ROM is the home page (see Figure 4.1). This page provides an overview of, as well as access to, the different sections of the CD-ROM. This page presents a concise summary of each of the main sections; to view each section, simply click on the hypertext link. Click on How to Use This CD-ROM for tutorials on selected tools.

When navigating the CD-ROM, remember that the main sections can be accessed in several ways: (1) clicking the navigation bar at the top of each page; (2) right-clicking on the screen and selecting a navigation bar item from the menu; and (3) using the home page hypertext links to each section.

Tutorials to help the user navigate the CD-ROM are located in the home section.

Note for Macintosh users:
Instead of right-clicking, use CTRL + Click to activate the contextual menu.

GLOSSARY

Downloadable materials — Materials that can be transferred from the CD to a personal computer

Grade level — Schooling level at which NAEP examinations are administered: Main NAEP at the 4th-, 8th-, and 12th-grade levels; State NAEP at the 4th- and 8th-grade levels

NAEP content strands — The five mathematics content areas used in the Main and State NAEP examinations: (a) Number Sense, Properties, and Operations; (b) Measurement; (c) Geometry and Spatial Sense; (d) Data Analysis, Statistics, and Probability; and (e) Algebra and Functions.

Many of the workshops use a set of student responses to NAEP test items as materials. To augment the student responses included in the materials, go to the NAEP Item Search and select the Item Number used in the workshop.

Figure 4.1. CD-ROM home page.

> **Companion CD**
>
> (X) NCTM NATIONAL COUNCIL OF TEACHERS OF MATHEMATICS
>
> **Learning From NAEP**
> Professional Development Materials for Teachers of Mathematics
>
> **Workshops**
> Access six fully developed workshops, and modify them to suit your needs.
>
> **NAEP Item Search**
> Explore more than one thousand student responses to 50 released NAEP items along with score guides and performance data.
>
> **Resources**
> Download State Snap Shot and State Gaps reports from the Nation's Report Card, select activities from the Activity Bank to supplement a workshop, and print references and glossary terms for each workshop.
>
> **About the IU-NAEP Project**
>
> CD-ROM Tutorial Movies
>
> Copyright Information

Workshops

The CD-ROM works in tandem with this manual to provide everything needed to present, participate in, and modify the workshops in chapters 5–10. Each workshop has a corresponding page on the CD-ROM, with downloadable materials and supporting resources for that workshop. To access those materials, click on Workshops in the navigation bar to bring up a table that lists and briefly describes all the workshops. The last two columns of the table offer two options: use the workshop exactly as it is from the manual—with the grade level and content area already chosen—or modify the workshop by choosing a different grade level and content area. Clicking on Modify brings up a dialogue box for selecting the desired Grade Level and Content Area (see Figure 4.2). For State NAEP you are asked to select a state. For convenience, the dialogue box shows only those choices that are applicable. If the workshop cannot be modified for a particular grade level or content area, the option will not be available. In addition, in this manual, the overview section of each workshop lists the grade level and content area for which the workshop can be modified.

Once the content and grade level of the workshop are chosen, a single workshop page appears. This page lists all the downloadable materials for the workshop in a table, as well as several resources above the table. For example, the materials table links to each document referenced in the manual through the use of a reference number that corresponds to the reference number used in the materials table in the manual. To help the user keep track of materials, the reference number contains the chapter number, the activity number, and the material number (see Figure 4.3). To download the materials, click on each material to download it individually.

Figure 4.2. CD-ROM workshops: Modify pop-up menu.

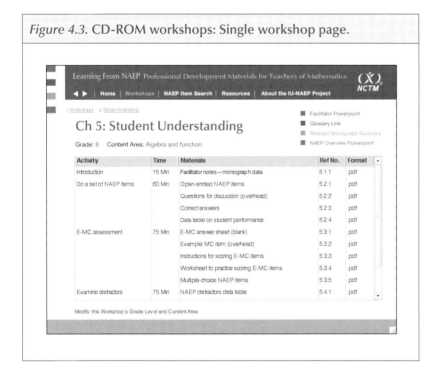

Figure 4.3. CD-ROM workshops: Single workshop page.

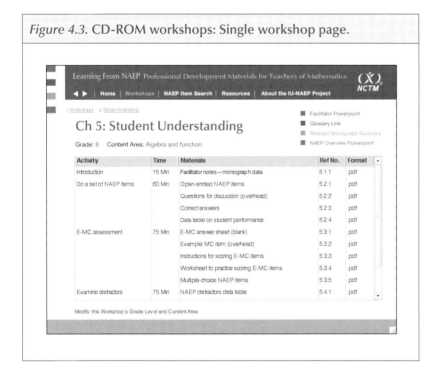

The single workshop page also has materials to help facilitate the workshop. These aids appear as icons above the materials table and include links to the following resources:

- Facilitator PowerPoint: A downloadable organizational device that provides a step-by-step guide through the workshop

- Glossary link: A link to the glossary entry for the workshop chapter in the Resources section; click on Print Window to download a printable version of the glossary

Several of the resources listed above the table on the single workshop page can also be found in the Resources section of the CD-ROM. Visit the Glossary for a complete list of terms used in all the workshops, and go to the Activity Bank and choose Activity 1: "Getting an Overview of NAEP" to access the script and discussion guide that accompany the NAEP Overview PowerPoint. All the chapter summaries from the monograph can also be accessed in the IU-NAEP section of the CD-ROM.

Appropriate software is required to view PDF documents. Acrobat Reader is available as a free download from Adobe at http://www.adobe.com/products/acrobat/readermain.html.

- Relevant monograph summary: A downloadable PDF file of a summary of the monograph chapter or chapters (from Kloosterman & Lester, 2004) that support the workshop

- NAEP Overview PowerPoint: A downloadable PowerPoint history of NAEP (additional information available in chapter 1 of this manual)

The materials on the CD are presented in several formats; the most common are PDF documents, text documents, Excel files, and PowerPoint presentations. Copies of the appropriate software must be installed on the computer being used to access these materials. When the user chooses a material or clicks on an icon from the CD, a document will open in its own application (e.g., a PDF file will open in Adobe Reader or Adobe Acrobat). When opened, the desired document can be printed or saved to the desktop.

NAEP Item Search

The NAEP Item Search on the CD-ROM is designed to access NAEP test items released by NCES; it is similar to the NAEP online Question Tool discussed in chapters 2 and 3. Its purpose is to provide access to more extensive NAEP data and student responses than are available in just the workshops. It can be used for creating or modifying NAEP workshops and for personal professional development. It allows educators to continue their own investigation of NAEP data and student responses. The CD-ROM NAEP Item Search also offers a simple and reliable way to access the NAEP test item, scoring guides, content classifications, performance data, statistics, and 20 to 30 student responses for each item.

To search for NAEP test items on the CD-ROM, begin by clicking on the navigation heading called NAEP Item Search to open the item-search screen (Figure 4.4). A user can locate NAEP test items on the CD in three ways. The first search method is by NAEP Item Number, which has seven digits and begins with the letter M. Type in an item number, and click the Go button. If that item is available in the CD, the information for that test item will appear in the single item page.

The advanced search lets the user narrow some or all of the search criteria without knowing the item numbers or which test items are available on the CD. In the advanced search, click Go to view a list of test items that fit the desired criteria. For a list of all NAEP test items available on the CD, simply click the corresponding Go button to view the list.

If the search has returned more than one item (using advanced search or View All Available NAEP Items), a Search Results page appears. This page features a table listing each NAEP test item that was returned by the search, along with the item number, grade, mathematical content area, and a short description of each item. The scrollbar to the right can be used to view all the items if too many are listed to display on one

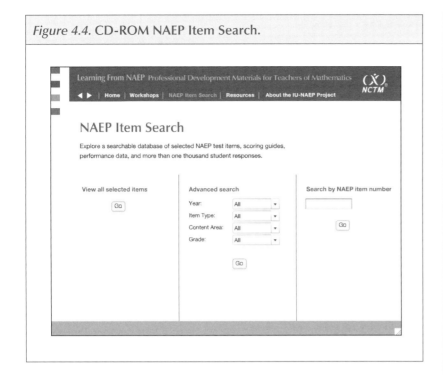

Figure 4.4. CD-ROM NAEP Item Search.

screen. Find the desired item, and click on its entry in the table to open the single item page.

The single item page displays detailed information about a NAEP test item. This single item page also allows the user to open PDF documents of the NAEP test question, scoring guide, student responses, and other data separately by checking the boxes next to them and clicking on the Open Selected button. Doing so will launch separate PDF documents for each item that is checked. To view all the data in one document, click on Open All (see Figure 4.5).

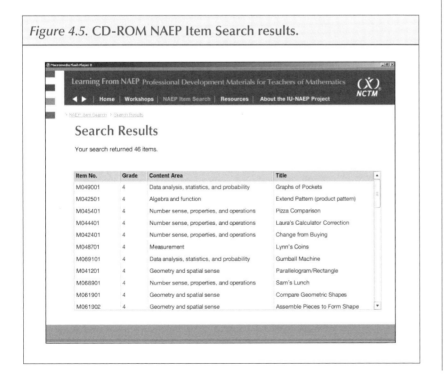

Figure 4.5. CD-ROM NAEP Item Search results.

GLOSSARY

Item — The basic scorable part of an assessment; a test question

Scoring guide/key — A list of the correct answers for multiple-choice questions; also, a holistic rubric used to score each student's response for short constructed-response and extended constructed-response items

Content classification — A designation of the mathematical content area(s) and knowledge that the selected item assesses

Performance data — Information that reveals the percentage of students at each score level

Student responses — Student answers that have been scored on the basis of the NAEP scoring rubric; not available for multiple-choice questions

NAEP item number — The number used to identify an individual NAEP item; typically begins with the letter M followed by six digits. Item numbers can be found on the More Data page, following the item description.

NAEP Questions Tool — The data available for items here mirror the tabs in the NAEP Questions Tool on the Web site (discussed in chapter 3):

— Question

— Performance Data

— Content Classification

— Scoring Guide

— Student Responses

— More Data

Although the CD-ROM Item Search offers more student responses than NAEP online Tools, it covers NAEP test items from 1992 and 1996 only. The most recent test items are available only online, as are graded responses.

Appendix A of the manual contains a description and overview of each activity in the Activity Bank.

Resources

The CD-ROM also provides resources, activities, materials, and student responses to create a new workshop that will meet specific needs or supplement an existing workshop from the manual. Opening this section brings up a list of subcategories with a short description of each resource section. Clicking on the hypertext heading leads to the desired area.

The Nation's Report Card

The National Assessment of Educational Progress (NAEP) is often referred to as the "nation's report card." Click on either State Snapshot Report or State Gaps Report. Documents for both kinds of reports are available for fourth and eighth grades and are listed alphabetically by state. Select the kind of report, and then select a state by scrolling. To download the grade-specific report, click on the appropriate hypertext.

Activity Bank

The Activity Bank accesses a set of generic activities for use as building blocks to create or modify a workshop. Clicking on Activity Bank opens a page with a table that lists the activities available, each with a short description and estimate of time required. Clicking on a table row brings up a dialog box from which to select the appropriate grade level and content strand, just as on the workshop pages. (If the grade or content area cannot be modified for an activity, the selection box will be gray.) Click Continue to advance to a single activity screen that contains a longer description of the activity, the estimated time, the chosen grade and content level, and a table with the materials needed for the activity (Figure 4.6). As with the workshop tables, click on each material to download it. Each activity has step-by-step instructions.

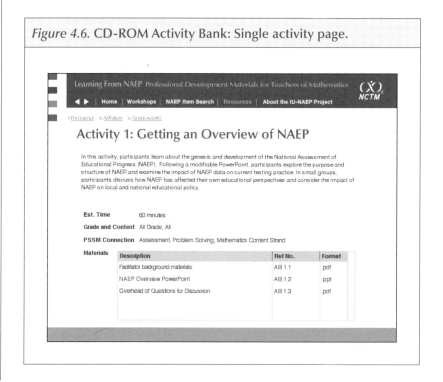

Figure 4.6. CD-ROM Activity Bank: Single activity page.

References

The References section contains a bibliography for each chapter that can be printed out and used as a resource during workshops. The references are arranged by chapter and are also available as a full-text list that can be modified.

Glossary

The glossary for a given chapter can be accessed in two ways: through the individual Workshop page or in the Resources section. In the Resources section, go to the glossary, choose the desired chapter from the manual, and print the window; or select All to print a complete list of terms with definitions. These terms are also defined in the margins of the manual.

About the IU-NAEP Project

The IU-NAEP section gives an overview of the IU-NAEP project, introduces the authors and developers of the *Learning From NAEP* materials, and summarizes each chapter in the monograph *Results and Interpretations of the 1990 through 2000 Mathematics Assessments of the National Assessment of Educational Progress* (Kloosterman & Lester, 2004).

The Workshops

Each of the next five chapters is a workshop that addresses a particular theme related to the analysis of NAEP data: student understanding, content knowledge, assessment, state issues, and equity. On the CD-ROM you will find downloadable versions of all the materials needed to implement the workshops. In addition, you can modify the workshops by grade level, by content, and in some cases, by state. Finally, you can use the Activity Bank on the CD-ROM and the reflections in Chapter 10 to construct your own workshops.

Diana V. Lambdin and Shelby P. Morge

THE FOCUS of this workshop is to help participants analyze mathematics test items to consider the thinking required to solve them. Participants explore open-ended items, multiple-choice items, and extended-multiple-choice items to investigate how the structure of test items affects how much students can show about what they know. Participants learn about national trends relating to a particular content strand and grade level as they examine test items and student performance data from the National Assessment of Educational Progress (NAEP).

Workshop Overview

Participants begin by working individually or in pairs to complete NAEP items. Initially, the items are worked as open-ended items (even if they were originally given in NAEP as multiple-choice items), and then participants discuss their answers and consider how their students might respond to these items. This approach gives participants firsthand experience in working the items and may lead to a discussion of common or anticipated errors. Participants engage in a large-group discussion of their findings, including common solutions and errors for each NAEP item.

> *A significant challenge for participants is to develop and rely on high-quality classroom assessments while ignoring the pressures of testing, often multiple-choice, that exist outside the classroom.*

Next, participants learn about the notion of extended-multiple-choice (E-MC) format—a test format that requires more thought from students, is easily scored, awards partial credit, and can offer educators more insight into student thinking than traditional multiple-choice examinations. Participants experience E-MC for themselves and then discuss the feasibility and potential benefits of using E-MC in classroom instruction and assessment.

 The following Principles and Process Standards, taken from NCTM's *Principles and Standards for School Mathematics* (2000), are emphasized in this workshop:

NCTM Principles
- ☐ Equity
- ☑ Teaching
- ☐ Learning
- ☑ Assessment
- ☐ Technology
- ☐ Curriculum

NCTM Process Standards
- ☑ Problem Solving
- ☐ Reasoning and Proof
- ☑ Communication
- ☐ Connections
- ☐ Representation

GLOSSARY

Open-ended item — A type of assessment item in which students must construct an answer rather than simply choose from among answer options provided. NAEP open-ended items come in several formats: items in which students simply write their answers in the space provided (short constructed-response, or SCR), items in which students answer multiple questions and provide a brief rationale for each response given (another sort of SCR), or items in which students provide extended constructed-response (ECR) answers.

Multiple-choice item — An item that consists of one or more introductory sentences followed by a list of response options that include the correct answer and several incorrect alternatives

Extended-multiple-choice item — A test item that requires more thought and response from students than a traditional multiple-choice item, is easily scored, awards partial credit, and can offer educators more insight into student thinking than an item in traditional multiple-choice format. E-MC can take many forms. In addition to picking an answer option from a list provided, students may be asked to indicate how sure they are about their choice, or to indicate which rejected options they know are wrong and which they are uncertain about, or to explain why the options they labeled wrong are incorrect, or to provide some other explanation of their thinking.

Distractor — An incorrect response choice included in a multiple-choice item

NAEP content strands — The five mathematics content areas used in the Main and State NAEP examinations: (a) Number Sense, Properties, and Operations; (b) Measurement; (c) Geometry and Spatial Sense; (d) Data Analysis, Statistics, and Probability; and (e) Algebra and Functions

 The following workshop provides specific examples from the eighth-grade Algebra strand. To modify the session for other grade levels and content areas, go to the Workshop section of the CD-ROM and click on Modify.

In small groups, participants carefully examine real distractors from the NAEP items and compare those distractors with their common solutions and errors. Participants attempt to predict which NAEP distractors will be popular or unpopular student responses and discuss why they think so. Finally, participants come together to discuss information presented in a data table: the actual percentage of students who selected each option during NAEP data collection. Participants discuss what makes good multiple-choice distractors as well as how best to use multiple-choice items in their classrooms. This workshop allows participants to learn more about student understanding as revealed by different types of test items.

Goals

- To highlight the importance of analyzing mathematics test items to determine the cognitive thinking required to solve them

- To help participants think about how the structure of test items affects how much students can show about what they know

- To introduce the E-MC test format and explore how E-MC may give participants insights into the thinking of students

- To inform participants about NAEP trends relating to content-strand instruction and grade-level student performance

Timing: 3 hours 45 minutes

Grade Band: ☑ 1–5 ☑ 6–8 ☑ 9–12

NAEP Content Strand

☑ Number Sense, Properties, and Operations

☑ Algebra and Functions

☑ Data Analysis, Statistics, and Probability

☑ Geometry and Spatial Sense

☑ Measurement

Materials

Activity	CD-ROM No.	Materials	Number Needed
1. Introduction (15 minutes)	5.1.1	Facilitator Notes—Monograph Data	1 per facilitator
2. Examine a Set of NAEP Test Items (60 minutes)	5.2.1	Open-ended NAEP Items	1 per participant
	5.2.2	Questions for Discussion (overhead)	1 per facilitator
	5.2.3	Correct Answers	1 per facilitator
	5.2.4	Data Table on Student Performance	1 per facilitator
3. Extended Multiple-Choice (E-MC) Items (75 minutes)	5.3.1	E-MC Answer Sheet (blank)	1 per facilitator
	5.3.2	Example of MC Item (overhead transparency)	1 per facilitator
	5.3.3	Instructions for Scoring E-MC Items	1 per participant
	5.3.4	Worksheet to Practice Scoring E-MC Items	1 per participant
	5.3.5	Multiple-Choice NAEP Items	1 per participan
4. Examine Distractors (75 minutes)	5.4.1	NAEP Distractors Data Table	1 per participant
	5.4.2	Distractor Questions (PowerPoint)	1 per facilitator
5. Wrap Up (15 minutes)		4 × 6 note cards	1 per participant

Background and Context Notes

A significant challenge for participants is to develop and rely on high-quality classroom assessments while ignoring the pressures of testing, often multiple-choice, that exist outside the classroom. For that reason, an important consideration is how test developers determine which multiple-choice distractors to include on assessments. That information can be helpful in developing classroom assessments that guide instructional decisions, with the goals of improving student learning and better preparing students for high-stakes assessments.

Several researchers have offered guidelines for writing appropriate and effective multiple-choice items (e.g., Haladyna, Downing, & Rodriguez, 2002; Haladyna, 2004) while explaining different reasons for deciding which distractors to include. The reasons may be statistical (e.g., to spread student scores) or pedagogical (e.g., to illuminate how students are thinking). Analyzing the distractors is helpful in determining their relative usefulness; any alternatives that most students consistently fail to select should probably be either modified or eliminated.

From the pedagogical perspective, creating a good multiple-choice test begins with written specifications that describe the skills and knowledge to be tested (Fuhrman 1996). Some educators recommend, however, that teachers should avoid using "preanswered" tests whenever possible. Van de Walle (2004, p. 73) explains that "tests of this type tend to frag-

To download the materials for this workshop, go to the Workshop page on the *Learning From NAEP* CD-ROM and click on Manual or Modify next to the title of the workshop. A facilitator can modify the workshop by grade level and content area or use the workshop as it is described in the manual.

When using several or all of the activities in this workshop, the same set of test items should be used with all activities.

GLOSSARY

High-stakes test — An assessment instrument used to make significant educational decisions about students, teachers, schools, or school districts

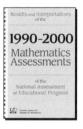

Data about student performance in various mathematics content strands, and at various grade levels, can be found in the monograph by Kloosterman & Lester (2004). That monograph contains three chapters about changes in students' knowledge of mathematics from 1990 to 2000: in Grade 4 (chapter 4), in Grade 8 (chapter 5), and in Grade 12 (chapter 6); as well as about Long-Term Trend NAEP, which examines student performance on a set of items held consistent since 1973 (chapter 7), and about student achievement on State NAEP (chapter 9).

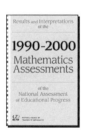

Chapter 12 of the monograph by Kloosterman & Lester (2004, pp. 337–362) discusses national data on students' NAEP performance from 1990 to 2000 in constructing responses to open-ended (SCR and ECR) items.

ment what children have learned and hide most of what they know." He emphasizes that appropriate assessment practices are integral to the instructional process and should focus on determining what students know instead of what they do not know. As a result of this workshop, participants will be better able to make informed decisions about the types of assessment to use in their classrooms.

Preparing to Facilitate

Before beginning, review the "Facilitator Notes—Monograph Data" (5.1.1) for the chosen content strand and grade level so that you can begin the workshop with a short statement presenting background information about recent trends in NAEP student performance in that content strand and at that grade level. Also review "Worksheet to Practice Scoring E-MC Test Items" (5.3.5) so you feel prepared to explain this procedure to participants. Review any other materials that might help familiarize you with the workshop format.

You will need to have one copy of all materials per facilitator and per participant, because some of the information will be used later in the workshop. Also, we suggest *not* giving participants the materials all at once (e.g., in a binder or folder), because some activities ask participants to predict answers to questions or to predict student performance data before those things are actually revealed in a subsequent activity.

Facilitating the Workshop

Activity 1: Introduction
Provide the context

Using the "Facilitator Background Notes—Monograph Data" (5.1.1) as a guide, take 10 minutes to explain the format of the workshop and to highlight the significance of working within the chosen mathematics content strand.

Discuss the data

Ask participants about their prior experiences with the mathematics content strand and grade level that they will be examining during this workshop. What common difficulties have their students had with that content? What common misconceptions are they aware of? What do the national data (from Kloosterman & Lester [2004] or from the NAEP Web site [http://nces.ed.gov/nationsreportcard/]) show about student performance in the chosen strand? Has student performance nationwide changed much in recent years? Do 4th graders exhibit different performance patterns than 8th or 12th graders do in that content strand? What might account for differences observed over time or by grade level?

Activity 2: Examine a Set of NAEP Test Items
Work the NAEP test items

Allow the participants approximately 20 minutes to work the test items. Items can be found in "Open-Ended NAEP Items" (5.2.1). Participants may work the items alone or in pairs. Walk around, and observe the participants as they work. Join in the problem solving and discussion going on at this time, without giving away the answers. Ask the participants such questions as these:

- Could you do the test item in a different way?

- How do you think your students might approach this test item?

- How is this test item different from others you have seen?

Organize small-group discussion

Organize the participants into small groups to discuss a few questions related to the test items they have just worked. They should spend approximately 20 minutes answering the questions. Divide up the items so that each group attempts to answer the questions that follow by critically examining at least two test items. The questions are provided in "Questions for Discussion" (5.2.2). Post the questions on an overhead transparency, or make enough copies so that each small group has a list to share.

- What does this NAEP item have the potential to reveal about students' understanding of the content strand?

- What mathematics is important in this item? What types of thinking would be important for success with this item or with similar items in this content strand? (As a follow-up to this question, ask what types of mathematics education goals the item could address. For example, the item may address concepts, factual information, skills and procedures, mathematical reasoning and proof, problem solving and applications, or confidence and competence.)

- If a student was explaining how to do this test item, what should a teacher expect to hear or read in a good oral or written explanation?

- What errors or misconceptions might be likely?

Participants' responses to these questions may vary, leading to a discussion about which items are more worthwhile for students as well as what each item reveals about student thinking. The discussion may center on mathematics standards or curricula that are implemented within

In the workshop in the manual, participants examine eighth-grade NAEP pattern items. Because student performance on these items has varied in the past, a close look at student performance and misconceptions is informative. Fourth-grade students have shown slight improvements on NAEP pattern items since 1990. However obvious differences have been noted between their performances on repeating-pattern items and growth-pattern items. A higher percentage of fourth-grade students correctly responded to repeating-pattern items in 1996 and 2000 (see Table 4.19). Similar results have been found with eighth-grade students. Their performance on making a prediction when observing a repeating pattern has increased, most notably when the pattern is composed of letters. In 1990 only half of the eighth-grade students tested could find the next letter in a pattern, but by 2000 that number had increased to almost two thirds. Making predictions in growth patterns was also more difficult for eighth-graders. However, the ability of students to perform well on these particular items has increased over the past decade (see Table 5.21). Note comparisons between fourth-grade and eighth-grade percentages of correct responses on similar NAEP items in Table 4.19. The gains in recognizing and completing pattern items may be a result of greater curricular emphasis on patterns and algebra. This increase in emphasis was reported by teachers in the NAEP teacher questionnaire. (Based on "Facilitator Notes—Monograph Data," 5.1.1)

 A generic version of all activities used in this workshop can be found on the CD-ROM in the Activity Bank and used to build a new workshop.

Workshop

Timing will vary depending on the number of NAEP items used in the workshop.

In response to these follow-up questions, participants may suggest such modifications as using different multiple-choice distractors, giving the assessment item as an open-ended item, or asking students to offer an explanation for their answer choices.

For the manual version of the workshop, which focuses on eighth-grade algebra, two open-ended NAEP items are included (Find the Next Figure in the Pattern, M022101, and Marcy's Dots, M054301) as well as several multiple-choice items for participants to work out in open-ended format. Modified versions of this workshop also incorporate multiple-choice and open-ended NAEP items. The open-ended items offer the opportunity for discussing how they can be made into multiple-choice items (by identifying common student responses to pose as distractors). And, of course, this workshop also shows how multiple-choice items can be made into open-ended items by simply removing the distractors.

the participants' school district. Participants may also consider the performance of students at other grade levels on similar items and compare it with performance at the grade level addressed in this workshop.

Have large group discuss small-group findings

After participants have had ample time to address the list of questions, pull them back together into a large group. Ask participants whether some test items seemed to be more difficult than others. Take a poll, and ask for predictions about which items will be the easiest and hardest for students. List the test items where everyone can see them and attempt to order them by perceived level of difficulty or to assign them to categories (e.g., easy, medium, hard, very challenging).

Then reveal only the percentages right and wrong for each NAEP item. The percentages for each item can be found on "Correct Answers" (5.2.3). If the items are multiple choice, do not yet share the distractors with workshop participants. Spend the remainder of the 20 minutes on highlights from each small-group discussion. Encourage participants to think about how well their predictions of difficulty level fit with the actual student performance data. For items that create significant discussion, ask the following questions:

- Why do you think this item was so challenging (or so easy) for students?

- If you were to use this item with your students, how might you modify it? Why?

- What mathematics education goals might be further addressed by your modifications?

Activity 3: Extended Multiple-Choice (E-MC) Items
Introduce E-MC

The extended multiple-choice (E-MC) format requires students not only to select an answer for each item from several choices provided (as in traditional multiple choice) but also to provide information about their reasoning in choosing that answer and about their confidence in the choice (Joyner & Bright, 2004). Distribute copies of "Extended Multiple-Choice Answer Sheet" (5.3.1). Show "Example Extended M-C Item" (5.3.2) on the overhead projector, and ask all participants to put X through any options they *know* are incorrect, to circle the answer they *think* is correct, and to circle either SURE or NOT SURE concerning their thinking about the answer they circled. Distribute "Instructions for Scoring Extended-Multiple-Choice (E-MC) Items" (5.3.3), and explain how E-MC scoring works. Reveal the correct answer, and have participants score their own work.

Experience E-MC

Distribute hard copies of a collection of "Multiple-Choice NAEP Items" (5.3.5). Allow participants time to do each item, recording their answer using the E-MC format on their "Extended Multiple-Choice Answer Sheet" (5.3.1). Share the correct answers for each item. The answers for each item can be found on "Correct Answers" (5.2.3). Have participants score their papers by using the E-MC scoring scheme.

Make sure that the percentage of NAEP students selecting each multiple-choice option is not revealed to participants at this point. Tell them only the percentages right and wrong for each item.

Discuss E-MC and student thinking

Engage participants in discussing what students might learn from doing E-MC items and what additional insights into student thinking a teacher might get from using E-MC items in small groups. The following "Questions for Discussion" (5.2.2.) are the same as the questions used in the previous activity. Ask participants to revisit the questions and to offer responses based on their new experiences with the E-MC items.

Questions for Discussion (5.2.2)

- What does this item have the potential to reveal about students' understanding of the topic?

- What mathematics is important in this item? What types of thinking would be important for success with this item or with similar items in this content strand? (As a follow-up, ask what types of mathematics education goals the item could address. For example, the item may address concepts, factual information, skills and procedures, mathematical reasoning and proof, problem solving and applications, or confidence and competence.)

- If a student is explaining how to do this problem, what should a teacher expect to hear or read in a good oral or written explanation? What errors or misconceptions might be likely?

 To include additional open-ended items in an activity or workshop, use the NAEP Item Search to search by grade band or content to find additional items for which student performance data are available.

For more practice in scoring, distribute "A Worksheet to Practice Scoring E-MC Test Items" (5.3.4) to all participants. Have participants score the answers given by each of five students. Compare scores. Resolve any discrepancies.

Workshop

If "E-MC" is the final workshop activity and is not linked with the "Examine Distractors" activity, the facilitator may choose to reveal NAEP student performance (percentage of students choosing each distractor) for each of the items that were worked by the participants—see 5.4.1, "NAEP Distractors Data Table."

A table showing percentages of NAEP students choosing each multiple-choice option can easily be created for any group of NAEP items by first going to the NAEP online Questions Tool at the NAEP Web site (http://nces.ed.gov/nationsreportcard/itmrls/NQT_StartSearch.asp). Go to Advanced Search, and select Mathematics. On the resulting screen, select the desired grade level, content area, and question type to bring up a list of NAEP items. Click on the name of the desired item to view the actual NAEP item. The More Data tab reveals the percentages of students who chose the correct answer and each of the various distractors for the item

Wrap up E-MC

Spend the remainder of the discussion time on highlights from each small group. For NAEP items that create significant discussion, possibly revisit the following questions:

- If you were to use this item with your students, how might you modify it? Why?

- What mathematics education goals might be further addressed by your modifications?

Activity 4: Examine Distractors
Revisit distractors

NAEP multiple-choice distractors were revealed in Activity 3. At this point, ask the participants to make predictions about which NAEP distractors were selected most often. Briefly discuss how the NAEP distractors may or may not relate to the difficulties participants encountered when doing the test items as extended or open-ended.

Compare distractors with student performance

Pass out the "NAEP Distractors Data Table" (5.4.1). Allow 15 minutes for participants to work in small groups to discuss how their predictions of students' choice of distractors compared with the actual student performance data. As participants make comparisons, they should consider the following questions found on their data table:

- Which distractors were selected most often? Why do you think so?

- Which distractors were chosen infrequently? Why do you think so?

- For some items, was an incorrect distractor more popular, overall, than the correct answer? If so, what do you think made that choice compelling for students?

- If you were using these items with your students, how might you modify the choices you made available?

- Can these items be structured in a way that might lead students to the correct thinking required to solve the item without telling them the answer?

- What does this information about the distribution of student responses tell us about student learning and understanding of these items? Do you notice any commonalities in student thinking by examining the distractors chosen by students?

Participants may suggest that certain distractors were chosen because they represented common errors for that particular item. Perhaps they discovered those errors in their own work on the item. Other distractors may be appealing to students because they are closely related to the numbers given in the original NAEP item. Participants may suggest different distractors and give reasons for using those distractors. They may also suggest a total modification of the problem into an extended multiple-choice or open-ended item. In addition, participants may notice that students demonstrate a low level of understanding of the concept being tested. Perhaps they will suspect that some students have only a limited understanding of the concept, so they may suggest giving students more experiences with that sort of item.

Have groups discuss data table

Pull participants back together for a large-group discussion of responses. Use the PowerPoint "Distractor Questions" (5.4.2) as participants report their small-group discussions of the data tables. Wrap up the activity by highlighting the idea that participants can use the percentages of correct answers on multiple-choice items to learn more about student understanding and to guide future classroom instruction.

Activity 5: Wrap Up

Take 15 minutes to wrap up the session. Engage participants in a discussion about building appropriate classroom assessments and what can be learned from student responses. Drawing on what participants have learned from this workshop, discuss their opinions about when and how to use multiple-choice items in the classroom. Also talk about the best approaches to developing and using such test items in their classrooms.

Pass out note cards, and ask participants to list three things they learned from this workshop—two things they plan to use from this workshop, and one thing they would like to know more about. Ask participants to leave the cards on their tables.

RESEARCH

Other reasons for selecting or not selecting the distractors might be related to the number of choices offered (Bruno & Dirkzwager, 1995; Crehan, Haladyna, & Brewer, 1993) as well as the order in which the distractors are listed in the NAEP item (Huntley & Welch, 1993).

Teachers who develop their own multiple-choice tests for classroom assessments may be interested the "Guidelines for Developing Multiple-Choice Items" suggested by Halayda (2004), who has done extensive research on developing and validating multiple-choice test items. His guidelines are organized by categories: content guidelines, style and format concerns, writing the stem, and writing choices.

Participants may discuss the misconceptions that are revealed through incorrect answer choices. Remember that one of the goals of classroom assessment, according to Assessment Standards for School Mathematics *(NCTM, 1995), is to find out what students know so as to guide decisions that are made in the classroom.*

REFERENCES

Bruno, J. E., & Dirkzwager, A. (1995). Determining the optimal number of alternatives to a multiple-choice test item: An information theoretic perspective. *Educational and Psychological Measurement, 55*(6), 959–966.

Crehan, K. D., Haladyna, T. M., & Brewer, B. W. (1993). Use of an inclusive option and the optimal number of options for multiple-choice items. *Educational and Psychological Measurement, 53*(Spring), 241–247.

Fuhrman, M. (1996). Developing good multiple-choice tests and test questions. *Journal of Geoscience Education 44* (4), 379–384.

Haladyna, T. M. (2004). *Developing and validating multiple-choice test items* (3rd ed.). Mahwah, NJ: Erlbaum.

Haladyna, T. M., Downing, S. M., & Rodriguez, M. C. (2002). A review of multiple-choice item-writing guidelines for classroom assessment. *Applied Measurement in Education, 15* (3), 309–333.

Huntley, R. M., & Welch, C. J. (1993). *Numerical answer options: Logical or random order?* Paper presented at the Annual Meeting of the American Educational Research Association, Atlanta, GA.

Joyner, J. M., & Bright, G. W. (2004). *Dynamic classroom assessment: Linking mathematical understanding to instruction in middle grades and high school.* Vernon Hills, IL: ETA/Cuisenaire.

Kehle, P., Wearne, D., Martin, W. G., Strutchens, M. E., & Warfield, J. (2004). What do 12-grade students know about mathematics? In P. Kloosterman & F. K. Lester, Jr. (eds.), *Results and interpretations of the 1990 through 2000 mathematics assessments of the National Assessment of Educational Progress* (pp. 145–174). Reston, VA: National Council of Teachers of Mathematics.

Kloosterman, P., & Lester, F. K., Jr., Eds. (2004). *Results and interpretations of the 1990–2000 mathematics assessments of the National Assessment of Educational Progress.* Reston, VA: National Council of Teachers of Mathematics.

Kloosterman, P., Warfield, J., Wearne, D., Koc, Y., Martin, W. G., & Strutchens, M. (2004). Fourth-grade students' knowledge of mathematics and perceptions of learning mathematics. In P. Kloosterman & F. K. Lester, Jr. (Eds.), *Results and interpretations of the 1990 through 2000 mathematics assessments of the National Assessment of Educational Progress* (pp. 71–103). Reston, VA: National Council of Teachers of Mathematics.

National Council of Teachers of Mathematics (NCTM). (1995). *Assessment standards for school mathematics.* Reston, VA: NCTM.

National Council of Teachers of Mathematics (NCTM). (2000). *Principles and standards for school mathematics.* Reston, VA: NCTM.

Sowder, J. T., Wearne, D., Martin, W., & Strutchens, M. (2004). What do 8th-grade students know about mathematics? Changes over a decade. In P. Kloosterman & F. K. Lester, Jr. (Eds.), *Results and interpretations of the 1990 through 2000 mathematics assessments of the National Assessment of Educational Progress* (pp. 105–144). Reston, VA: National Council of Teachers of Mathematics.

Van de Walle, J. (2004). *Elementary and middle school mathematics: Teaching developmentally* (5th ed.). Boston: Pearson Education.

Christine Oster

THE GOAL of this workshop is to engage participants in exploring the mathematics and mathematical thinking related to a specific mathematics content area at a given grade level. The workshop also encourages participants to view students' responses to items not only as evidence of student knowledge but also as clues about how to direct or modify instruction.

Workshop Overview

The workshop begins with participants' completing a focus task that engages them in the mathematical content area of the workshop and that is more challenging than most NAEP items are. Participants discuss their answers, the various approaches to solving the task, and the thinking involved in their solutions. This discussion gives participants firsthand experience in responding to an open-ended task and focuses their thinking on the selected content area for the workshop.

> *Often teachers are not aware of the wealth of information that can be found in student responses to test items. Through engaging teachers in examining items and student responses, facilitators can "create opportunities, through interaction and discussion among teachers, for addressing the challenges and developing judgment ..." (Driscoll & Bryant, 1998, p. 29).*

Participants then respond to one NAEP item related to the mathematical content of the focus task. The ensuing discussion focuses on participants' solutions to the item and other ways in which students might respond. Next, participants examine a collection of student work samples for the same NAEP item. They classify the student responses on a continuum from exemplary to unsatisfactory on the basis of the evidence presented in the written answers. Participants select four pieces of student work, each representing a different section of the continuum. They consider the evidence presented in each solution and make conjec-

The following Principles and Process Standards, taken from NCTM's *Principles and Standards for School Mathematics* (2000), are emphasized in this workshop:

NCTM Principles
- ☐ Equity
- ☑ Teaching
- ☑ Learning
- ☑ Assessment
- ☐ Technology
- ☑ Curriculum

NCTM Process Standards
- ☑ Problem Solving
- ☐ Reasoning and Proof
- ☐ Communication
- ☐ Connections
- ☐ Representation

GLOSSARY

Evidence — Data from student work, including writing, representations, equations, drawings, and computations, that can be used to make conjectures

Conjecture — A hypothesis or assertion about what a student does or does not understand, based on the evidence in her or his work; a guess, supported by evidence, about what a student is thinking

Curricular implication — Actions to be taken to further a student's understanding, based on evidence and conjectures made from examining the student's responses

 The following workshop provides specific examples from the fourth-grade Number Sense strand with a focus on fractions. To modify the session for other grade levels and content areas, go to the Workshop section of the CD-ROM and click on Modify.

This workshop works well in a full-day workshop model as well as split into two to four shorter segments.

tures about each student's knowledge. Participants then discuss curricular implications—that is, actions to be taken to further students' understanding—based on the evidence and conjectures made from examining the student work samples.

Finally, participants discuss the mathematical knowledge that students must have to correctly respond to the selected NAEP item. They consider the fact that evidence of knowledge is embedded in both correct and incorrect responses. Participants also discuss how student work might be used not only as assessment but also as a point for beginning student instruction.

Goals

*   To help participants learn to examine the mathematical content of items and the evidence of mathematical knowledge that responses to items contain

*   To help participants consider how they might use student responses to items not only as a means of assessing students but also as a tool for guiding instructional practices

Timing: Between 4 and 5 hours

Grade Band: ☑ **1–5**   ☑ **6–8**   ☑ **9–12**

NAEP Content Strand

☑ Number Sense, Properties, and Operations

☑ Algebra and Functions

☑ Data Analysis, Statistics, and Probability

☑ Geometry and Spatial Sense

☑ Measurement

Materials

Activity	CD-ROM No.	Materials	Number Needed
1. Introduction (20–40 minutes) (optional)	6.1.1	Introduction to NAEP PowerPoint	1 per facilitator
2. Do the Focus Task (30 minutes)	6.2.1 6.2.2 6.2.3	Facilitator's PowerPoint Focus task Facilitator notes for the focus task Blank overhead transparencies and pens	1 per facilitator 1 per participant 1 per facilitator
3. Do the NAEP Test Item (30 minutes)	6.3.1	NAEP test item	1 per participant
National Performance Results (optional) (10 minutes)	6.3.2	Overhead of data from Nation's Report Card www.nces.ed.gov/nationsreportcard/), available on the CD-ROM	1 per facilitator
4. Examine Student Work (60–90 minutes)	6.4.1	Student responses in packet Student responses overhead transparency	1 per participant 1 per facilitator
5. Make Decisions about Student Understanding and Performance (60–90 minutes)	6.5.1	Handout, "Moving Toward Understanding"	1 per participant; 1 overhead transparency for facilitator
6. Wrap Up (20 minutes)			1 per participant

Background and Context Notes

The focus task is designed to help participants experience a mathematical task from a learner's perspective. As they do the task, participants should consider how students might approach it. This process may include uncovering the content knowledge necessary to solve the task and can help participants assess their own students' performance.

Often teachers are not aware of the wealth of information that can be found in student responses to test items. Through engaging teachers in examining items and student responses, facilitators can "create opportunities, through interaction and discussion among teachers, for addressing the challenges and developing judgment … judgment about the quality of mathematics in tasks, judgment about the appropriateness of tasks, judgment about the quality of student responses, and judgment about consequent actions" (Driscoll & Bryant 1998, p. 29).

This workshop is designed to help participants examine students' responses to open-ended test items to determine what students know and understand. It will also help participants consider how this information can be used to make informed decisions about the next appropriate instructional activities for individual students. The culmination of this

To download the materials for this workshop, go to the Workshop page on the *Learning From NAEP* CD-ROM and click on Modify or Manual next to the title of this workshop.

GLOSSARY

Focus task—A task to be explored by workshop participants to immerse them in the content of a given NAEP item before they solve the NAEP item itself. Such a task is intended to help participants experience some of the struggle that a student might encounter when solving a NAEP item.

RESEARCH

"Asking what abilities are required of students to solve a problem or what the problem contributes to our knowledge of student abilities focuses our attention on identifying performance criteria" (Moon, 1997, p. 40).

RESEARCH

"Assessment should support the learning of important mathematics and furnish useful information to both teachers and students. Assessment should be more than merely a test at the end of instruction to gauge learning. It should be an integral part of instruction that guides teachers and enhances students' learning" (NCTM, 2000b).

Information about NAEP and student performance can be found in *Results and Interpretations of the 1990 Through 2000 Mathematics Assessments of the National Assessment of Educational Progress* (Kloosterman & Lester, 2004, chapters 3–5) as well as the NAEP Web site (www.nces.ed.gov/nationsreportcard/).

To download a discussion guide for the Introduction to NAEP PowerPoint, go to the Activity Bank in the Resources section of the CD-ROM and choose Activity 1: "Getting an Overview of NAEP."

During a field test of this workshop, one facilitator noted that the discussion following the NAEP Overview PowerPoint went far beyond the time allotted because standardized testing is "such a controversial subject." The facilitator may want to modify the workshop by using the NAEP PowerPoint as the last activity.

workshop focuses on those instructional actions that the teacher might take, also referred to in the workshop as *curricular implications.*

Facilitating the Workshop

This particular workshop is written to involve participants in an investigation of fractions, a topic that is embedded in the content area Number Sense, Properties, and Operations. The main mathematical focus is on the question "What is the whole?"

Activity 1: Introduction

As the facilitator, you may want to introduce the workshop by using the PowerPoint provided (6.1.1) to present a brief overview of NAEP and build a context for the NAEP problem and student work samples used in the workshop. The Introduction to NAEP PowerPoint (6.1.1) can be edited to fit your needs or be merged with the Facilitator PowerPoint also provided on the CD (6.2.1).

Activity 2: The focus task
Introduce the rationale

Take 5 minutes to give the rationale for the activity. Take time to highlight the significance of working within the chosen mathematical content strand. You may have specific reasons for selecting a certain strand, such as to increase participants' awareness of content knowledge as a systemwide effort or to improve test scores in a particular content area.

Do the focus task

Distribute a copy of the "Focus Task" (6.2.2) to each participant, and display it on an overhead transparency for recording solutions and comments. Allow participants sufficient time to work on the task alone, in pairs, or in small groups. Because one outcome of the focus task is disequilibrium, give participants permission to vent their frustrations. Remind them that the ambiguity and difficulty of the task are supposed to be prompts for critical thinking. Be sure to work through the focus task independently before facilitating the workshop. Many of the focus tasks are challenging, but the CD contains "Facilitator Notes" for each focus task (6.2.3) that can help you when working with workshop participants.

Share solutions

Ask whether any participant is willing to share her or his answer with the group. Participants' answers should be recorded so that all of them can be viewed by the group. Encourage participants to justify their answers and to discuss similarities and differences in approaches to the focus task. Continue the discussion until all willing voices have been heard.

Discuss in small groups

Ask participants to reflect on their focus-task experiences in groups of three or four. Begin by posing a question, such as one of the following examples, that will help focus the participants' thinking on the content emphasis of the workshop:

- What mathematics does this task address?

- What did you learn from doing this task?

- What errors or misconceptions could lead to different conclusions?

- What do you need to understand to be able to do this task?

The final question prompts the beginning of a discussion that will continue through the workshop. At this point, participants may offer broad or vague responses. After examining student work and discussing what can be gleaned from examining student responses, participants will become more articulate about the understanding that students may or may not have when they approach the NAEP item. One approach might be to take a few minutes to share the highlights from each small-group discussion.

Wrap up

Engage the participants in a discussion about mathematical content by asking one or more of the following questions:

- What was the mathematical focus of the problem we just solved?

- What could students learn from doing this focus task as an assessment item?

- What might students learn from sharing answers and different ways of doing an assessment item?

Activity 3: The NAEP test item
Do the NAEP item

Pass out a copy of the "NAEP Test Item" (6.3.1) to each participant. Encourage the participants to solve the problem at least two ways, working individually or in pairs. Some participants may need more time than others. Those who finish before others may explore how they think a child might solve this problem.

"The first time I presented a focus task, I was in for a huge surprise. It made many people uncomfortable because the problem did not have a straight-forward answer that they could find quickly. As I reflected on the experience, I realized that the purpose of the focus task is precisely to make participants uncomfortable—to put them into a state of disequilibrium. Not only does that help them understand the experience the students have when confronted with a problem that is out of their range of experience, but it also helps them focus on the concept at hand."

RESEARCH

"A problem is, by definition, a situation that causes disequilibrium and perplexity" (Lester & Charles, 2003, p. 7).

The facilitator should maintain the role of a "questioner" and not a supplier or verifier of correct answers.

GLOSSARY

Disequilibrium — A state at which a person has reached the limits of his or her knowledge or has encountered a situation that does not fit into his or her conceptual framework. Further inquiry and experience can resolve this uncomfortable state by helping the person understand the problem or concept and perceive that it "makes sense."

Although all the questions posed in the workshop are contained on the Facilitator's PowerPoint, facilitators sometimes find that putting them on an overhead transparency makes them easier to use in the workshop.

Share solutions

As with the focus task, ask whether anyone is willing to share his or her solution with the group. A helpful tactic is to record participants' answers so that all can be viewed by the group. Encourage participants to justify their answers and to discuss similarities and differences in approaches to the item. Continue the discussion until all willing voices have been heard.

Discuss in small groups

The following questions are included on the facilitator PowerPoint (6.2.1). The facilitator may want to edit them to suit a particular purpose or make an overhead transparency to use during the workshop.

- What mathematics is important in this test item?

- How might a student approach this test item?

- What student errors or misconceptions might be likely?

- What mathematical knowledge is necessary to solve this problem?

- What might this test item reveal about students' understanding?

Share the highlights from each small-group discussion, focusing on the questions used.

Wrap up

As a wrap up, continue the discussion begun during the focus task: What do students need to understand in order to do this NAEP item? As participants consider more deeply what mathematical knowledge is necessary to understand a concept such as "what is the whole?" they will begin to consider the depth of knowledge necessary for students to be successful on this NAEP item. Participants may also want to consider the following questions:

- What could students learn from doing this test item?

- What might students learn from sharing answers and different ways of doing the test item?

- What similarities and differences do you see between the NAEP test item and the focus task?

You may also want to share additional information about the selected item available on the NAEP Web site (see chapter 3, "Exploring the NAEP Online Tools") or the CD-ROM NAEP Item Search for this problem. The Performance Data include the percentage of students who correctly answered the question (6.3.2).

Activity 4: Examine student work
Discuss the student work samples

Activities 4 and 5 are best done in small groups of three or four participants. If participants are not already grouped this way, form groups first and then distribute the samples of student work (6.4.1) to each participant. Ask the participants to discuss the student work. The following questions may help facilitate group discussion. An overhead transparency of these questions can also be used to focus the groups' attention on specific attributes of the students' work.

- What do you think the students knew or did not know that made their answers different from ours?

- Are you surprised by any of the answers? What aspects of the student responses surprise you?

- How many "correct" solutions have you found?

- Are some solutions "more correct" than others? What makes them so?

Focus participants' thinking on the potential revealed in student responses by posing the following question:

- Rather than think in terms of "correct" and "incorrect," is it possible to think of responses as "reasonable" and "unreasonable"?

As participants examine student work more closely, they will begin to see different layers of understanding not evident at first glance. This discovery starts them on the path of viewing student work with an eye toward teaching rather than grading. You might initiate a follow-up discussion by asking, "How do these two approaches—correct versus incorrect and reasonable versus unreasonable—compare when evaluating student work?" Bring the groups back together to briefly discuss their findings.

Find evidence of student understanding

Ask participants to find evidence of students' understanding. Participants may initially make such conjectures as, "The child does not know how to divide the pizza into equal halves." Continue to ask questions to help participants focus on the things students have written on the paper. The goal is to have the group form a common definition of *evidence*, similar to this one: "Evidence is something concrete actually seen in the student's work. It includes writing, representations, equations, drawings, and computations. It is data from student work that can be used to make conjectures."

Select one work sample. Ask participants to make a list of evidence they find. Participants will share evidence drawn from the work sample.

"It seems that introducing the NAEP rubric into this workshop distracts from the workshop purpose. A goal of this workshop is to help participants look at student work and discern what students do and do not understand. You want to move participants away from thinking about student work in terms of 'right' or 'wrong,' to viewing student errors as opportunities for instruction."

"After we had done both the focus task and the NAEP problem, a participant remarked, 'I'm glad we did that hard problem first. I had a much better perspective when I did the NAEP problem because of that experience.' "

Record evidence on an overhead transparency or chart paper so that all participants can see it. Continue until all participants clearly understand the concept of evidence.

Make conjectures about student understanding

Using the same questioning procedure discussed for evidence, help the participants define *conjecture:* A conjecture is a hypothesis or assertion about what a student does or does not understand based on the evidence in his or her work. It is a guess, supported by evidence about what a student is thinking. Ask the participants to go back and look at the work sample they used to collect evidence. Encourage the participants to make conjectures based on the evidence they found.

Activity 5: Make decisions about student understanding and performance
Sort student work

Ask the participants to order the student work samples on a continuum from exemplary to unsatisfactory responses. They may choose to work individually or with a partner within their group. When participants later select four student work samples to analyze, this sort will help them choose four very different pieces of work if they pick one sample from each quarter of the continuum.

Encourage participants to discuss why they have sorted the work as they have. They do not have to agree, but they should justify their decisions to their group. In the collection of student work are examples of responses in which the answers are correct but the explanations show that the students did not understand the problem or concepts. Also included are examples of responses in which the answers are incorrect but the evidence indicates that the students have used at least some correct thinking. While circulating the room, ask whether participants are making conjectures or looking at the evidence, to help participants become more familiar with those terms.

Look for evidence, and make conjectures

Review the definition of *evidence.* It is something concrete that an observer can see in a student's work. Review also that a conjecture is a hypothesis, based on the evidence, about what a student does or does not understand.

Focus the participants' attention on one student work sample, and model recording the evidence on an overhead transparency of the first page of the worksheet "Moving Toward Understanding" (6.5.1) as the participants discuss the sample. Continue modeling, filling in the worksheet and taking suggestions from the participants for conjectures after the evidence has been recorded.

Discuss curricular implications

Have each small group brainstorm the next actions that the participant, as a teacher, could take to further the student's understanding. Those instructional modifications constitute the curricular implications for the student. Discuss participants' ideas, and continue modeling, filling in the worksheet "Moving Toward Understanding" on the overhead transparency.

At this point, the participants must draw on their experiences to make recommendations for the next step in teaching a child. For example, in this workshop, the concept of the whole is an idea that students learn through many varied experiences with fractions. Experiences that invite students to explore fractions using such materials as fraction strips, counters, or fraction circles will help students view fractions in a variety of settings so that they can easily transfer their knowledge to the pizza in the NAEP problem. Comparing fractional parts of different groups, such as pairs of shoes or the amount of seating available in a room, will help students build understanding about the fact that the whole can be represented in many different ways. The more varied experiences students have, the more understanding they will bring to solving problems. The process of devising curricular implications moves the focus of the NAEP test item from being an assessment to becoming a tool for instruction.

Examine multiple student work samples

Next pass out a copy of worksheet 6.5.1, "Moving Toward Understanding," to each participant. Request that the participants select four samples of student work that they find particularly interesting. If participants have sorted the student work on a continuum, they can easily select one piece from each quarter of their sort and have four very different samples to examine. If the sorting activity was omitted, ask participants to select four pieces of work showing various levels of proficiency. Participants should focus on each piece of student work on the worksheet. They should include the problem number and list the evidence found in the student work. They then make conjectures about the knowledge that the child has and the mathematical content that the child understands. This understanding can include thinking that has potential but that has not been explored or developed by the student. Finally, the participants identify curricular implications by listing the knowledge the child needs to learn and what activities or questions might extend this student's understanding of the concept.

Encourage participants to examine the student work as if they were the classroom teacher. What should they do next? Encourage participants to identify the edge, or periphery, of the student's understanding— the point at which instruction is most advantageous to the student's learning—and to articulate what that instruction might be.

RESEARCH

"Teacher groups can and should learn to analyze a range of student work on a task and to determine points where students need further instruction" (Driscoll & Bryant, 1998, p. 18).

Workshop

RESEARCH

"Is it possible for a student to perform at a low level on a test item while demonstrating thinking that shows great potential?" (Driscoll & Bryant, 1998, p. 17)

Activity 6: Wrap Up
Discuss the events of the workshop

Help the participants bring together the different strands in this workshop: an awareness of the mathematical concept under discussion, the fact that numerous correct paths can be taken to the right answer, the fact that examining evidence in students' work enables the teacher to make conjectures about what they do and do not know, and the identification of instructional steps that would benefit the student.

The following questions found on the PowerPoint might help the participants reflect on these ideas:

- Why might this NAEP item be difficult for students?

- What can we learn by looking at students' work?

- What you have learned about teaching fractions?

- What have you discovered about students' understanding of fractions?

- What can you do to help students take the next step in their learning?

Reflect on the workshop

Have participants write a brief statement on an index card about what impact this experience will have on their teaching. Use the following questions on the PowerPoint as prompts:

- What have you discovered about the curricular topic under discussion?

- How might you find out about a child's understanding?

- How are assessment and instruction related?

Encourage participants to share their statements.

REFERENCES

Driscoll, M., & Bryant, D. S. (1998). *Learning about assessment, learning through assessment.* Washington, DC: National Academy Press.

Kloosterman, P., and Lester, F. K., Jr. (Eds.). (2004). *Results and interpretations of the 1990 through 2000 mathematics assessments of the National Assessment of Educational Progress.* Reston, VA: National Council of Teachers of Mathematics.

Kloosterman, P., Warfield, J., Wearne, D., Koc, Y., Martin, W. G., & Strutchens, M. (2004). Fourth-grade students' knowledge of mathematics and perceptions of learning mathematics. In P. Kloosterman & F. K. Lester, Jr. (Eds.), *Results and interpretations of the 1990 through 2000 mathematics assessments of the National Assessment of Educational Progress* (pp. 71–103). Reston, VA: National Council of Teachers of Mathematics.

Lester, F. K., & Charles, R. I. (2003). *Teaching mathematics through problem solving, prekindergarten–grade 6.* Reston, VA: National Council of Teachers of Mathematics.

Moon, J. (1997). *Developing judgment: Assessing children's work in mathematics.* Portsmouth, NH: Heinemann**.**

National Council of Teachers of Mathematics (NCTM). (2000a). *Principles and standards for school mathematics.* Reston, VA: NCTM.

National Council of Teachers of Mathematics (NCTM). (2000b). *Principles and standards for school mathematics: An overview.*. Reston, VA: NCTM.

Richardson, K. (1997). *Math time, thinking with numbers video guide.* Norman, OK: Educational Enrichment.

Richardson, K. (2003). *Thinking with numbers workshop.* Bellingham, WA: Math Perspectives.

Kathleen Lynch-Davis and Fran Arbaugh

THIS workshop is designed to be implemented in three consecutive sessions lasting 2 to 3 hours each: (1) Examining Assessment Items, (2) Judging the Quality of Student Responses, and (3) Gathering Evidence. Participants analyze NAEP items and student work and create and use holistic, descriptive, and analytic rubrics. By engaging in this workshop, participants learn about making instructional decisions based on assessment practices.

Workshop Overview

Participants begin Session 1 by analyzing a set of 20 NAEP items using a particular framework that focuses on characteristics of assessment items. After completing that analysis, participants then sort the same set of items according to the types of student thinking that each item elicits. At the end of this session, participants reflect on the activities and what they have learned.

> *By examining the four phases of assessment within the context of student work, mathematics teachers can consider what types of items elicit what types of student thinking, how to assess student thinking, and how to make instructional decisions based on what their students know and can do.*

Session 2 focuses on judging the quality of student responses. Participants engage in activities that focus on rubric development and rubric analysis using a specific NAEP item and the associated student work. After completing the item, participants sort student work and create their own rubrics. They are then able to examine the rubric provided by NAEP for that item. At the conclusion of this session, participants reflect on the usefulness of the two rubrics.

 The following Principles and Process Standards, taken from NCTM's *Principles and Standards for School Mathematics* (2000), are emphasized in this workshop:

NCTM Principles
- ☐ Equity
- ☑ Teaching
- ☑ Learning
- ☑ Assessment
- ☐ Technology
- ☑ Curriculum

NCTM Process Standards
- ☑ Problem Solving
- ☐ Reasoning and Proof
- ☐ Communication
- ☐ Connections
- ☐ Representation

RESEARCH

In a review of 250 research studies on classroom assessment, Black & Wiliam (1998) concluded that classroom-based formative assessment, when appropriately used, can positively affect learning.

GLOSSARY

Item — The basic, scorable part of an assessment; a test question

Evidence — Data from student work, including writing, representations, equations, drawings, and computations, that can be used to make conjectures

Holistic rubric — A scoring scheme that assigns an overall level of performance by assessing performance across multiple criteria

Descriptive rubric — A scoring scheme that creates broad categories, such as setting up the problem, and scores each area separately (Moon, 1997)

Analytic rubric — A scoring scheme that articulates levels of performance for each criterion so the teacher can assess student performance on each criterion

RESEARCH

"Teachers must have tools and other supports if they are to implement high-quality assessment practices and use the resulting information to inform instruction" (Pellegrino, Chedowsky, & Glaser, 2001).

The following workshop does not provide specific examples from a particular grade level or content strand. The workshop can be modified by grade level; the content focus for each grade level includes all the strands.

"Assessment should support the learning of important mathematics and furnish useful information to both teachers and students" (NCTM, 2000, p. 22).

In Session 3, participants examine their own students' work on a particular NAEP item through the use of a rubric they develop as well as through the use of the NAEP scoring rubric.

Goals

- To support participants in making decisions with a goal of improving their instruction: plan for assessment, gather evidence using assessment items, interpret that evidence using rubrics, and use the resulting information to shape their instruction and future assessments

- To help participants think about how the characteristics of assessment items influence student responses

Timing: 2–3 hours per session (6–9 hours total)

Grade Band: ☑ 1–5 ☑ 6–8 ☑ 9–12

NAEP Content Strand

☑ Number Sense, Properties, and Operations

☑ Algebra and Functions

☑ Data Analysis, Statistics, and Probability

☑ Geometry and Spatial Sense

☑ Measurement

Materials

Activity	CD-ROM No.	Materials	Number Needed
1. Examine Assessment Items (2.5 hours)	7.1.1	Item set	1 per pair
	7.1.2	"Analyzing Assessment Items Framework"	1 per pair
	7.1.3	Item-Analysis Recording Sheet	1 per pair
	7.1.4	Item-Sort Recording Sheet	1 per facilitator
		Overhead transparency of the recording sheet (7.1.4)	1 per facilitator
		Overhead transparency of item set (7.1.1)	1 per facilitator
	7.1.5	Writing prompt (also on Facilitator PowerPoint)	1 per facilitator
2. Judge the Quality of Student Responses (3 hours)	7.2.1	Overhead transparency of item option	1 per facilitator
	7.2.2	Student work packets	1 packet per 2–3
		Chart paper	1 piece per 2–3
	7.2.3	Copy of NAEP rubric and student samples	1 packet per 2–3
	7.2.4	Overhead transparency of the comparison sheet	1 packet per 2–3
	7.2.5	Overhead transparency of student performance data	1 per facilitator
	7.2.6	Overhead transparency of extended results (More Data)	1 per facilitator
	7.2.7	Writing Prompt (also on Facilitator PowerPoint)	1 per facilitator
4. Gather Evidence (90 minutes over two meetings)		NAEP item from the item set (7.2.1)	1 per participant
	7.3.1	Writing Prompt (also on Facilitator PowerPoint)	1 per facilitator

Background and Context Notes

Issues of assessment are wide-ranging and complex. In a review of 250 research studies on classroom assessment, Black & Wiliam (1998) concluded that classroom-based formative assessment, when appropriately used, can positively affect learning. They argue that if mathematics teachers focus their efforts on classroom assessment that is primarily formative in nature, students' nationwide learning gains would have the potential to move the United States from an "average" country into the "top five" on The International Mathematics and Science Study (TIMSS). However, research has indicated that effective use of formative assessment practices is rare, and classroom assessment is generally poor (e.g., Cooney, Badger, & Wilson, 1993). In a national study of 364 mathematics and science teachers in Grades K–12 (Weiss, Pasley, Smith, Banilower, & Heck, 2003, pp. 104–105), the researchers found that the most common formative assessment practice is the use of "low-level 'fill-in-the-blank' questions, asked in rapid-fire, staccato fashion with an emphasis on getting the right answer and moving on, rather than helping the students make sense of ... concepts." Clearly, teachers

 To download the materials for this workshop, go to the Workshop page on the *Learning From NAEP* CD-ROM and click on Modify or Manual next to the title of this workshop.

 For the user's convenience, the Writing Prompts at the end of each activity (7.1.5; 7.2.7; 7.3.1) have been provided both as downloadable PDF files that can be made into overhead transparencies and also as part of the Facilitator PowerPoint. Facilitators can choose the format that better suits their needs.

Workshop

GLOSSARY

Formative assessment — A feedback process that furnishes information that can be used to fine-tune or modify an existing instructional approach

Assessment Standards for School Mathematics (NCTM, 1995) lists four goals of assessment: (1) monitoring student progress, (2) evaluating students' achievement, (3) evaluating programs, and (4) making instructional decisions.

must have appropriate tools and other supports if they are to implement high-quality assessment practices and use the resulting information to make informed instructional decisions (Pellegrino, Chedowsky, & Glaser, 2001). Adopting these instructional practices necessitates effective professional development focused on improving assessment.

This workshop is designed to support participants' use of classroom assessment to make instructional decisions, with a goal of improving instruction. Figure 7.1 contains this framework for thinking about using assessment to improve instruction (NCTM, 1995, p. 27). Assessments involve four phases: (1) plan the assessment, (2) gather evidence, (3) interpret the evidence, and (4) use the results. By examining each of these phases within the context of student work, participants can consider what types of items elicit what types of student thinking, how to assess student thinking, and how to make instructional decisions based on what their students know and can do. This workshop is designed to engage participants in considering all four phases of assessment.

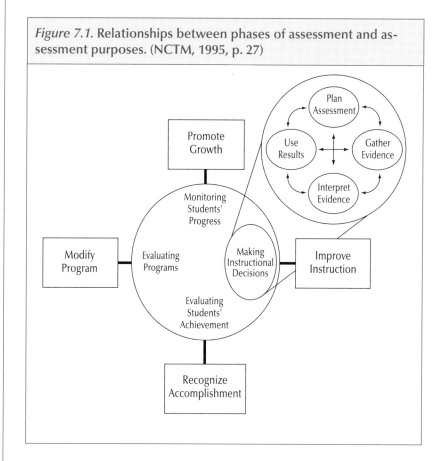

Figure 7.1. Relationships between phases of assessment and assessment purposes. (NCTM, 1995, p. 27)

Preparing to Facilitate

Before the workshop, we suggest that facilitators of this workshop become familiar with the National Council of Teachers of Mathematics (NCTM) document *Assessment Standards for School Mathematics* (Reston, VA: NCTM, 1995). Facilitators should also read the article "Improving Classroom Tests as a Means of Improving Assessment" by Denisse Thompson, Charlene E. Beckmann, and Sharon Senk (January 1997) (*Mathematics Teacher, 90,* 58–64). The article can be found under References in the Resources section of the CD-ROM.

When preparing materials for the workshop, keep in mind that many of the materials are designed for groups of two or three participants. In addition, many of the overhead transparencies listed in the materials can also be presented as PowerPoint slides and inserted into the Facilitator's PowerPoint.

Activity 2 offers several options for test items with student responses. The single workshop page in the Workshop section of the CD-ROM offers choices of a grade level and an option. These options refer to specific test items and accompanying student responses. To select a different item, use the back button to return to the Workshop Overview page and select the workshop again, choosing a different option. Please note that Activity 1 uses a different set of test items.

Facilitating the Workshop

ACTIVITY 1: Examine Assessment Items

Participants work in pairs or small groups to analyze a set of NAEP assessment items, using a specific framework. Participants next engage in an item-sorting activity in which the focus is to use a framework that focuses on the types of student thinking that different items elicit. This session offers participants the opportunity to think deeply about characteristics of assessment items and to develop a common language for characteristics of assessment items.

Introduction

Take 5 minutes to introduce the session. The focus of this brief introduction is to give participants a little bit of background about the importance of thinking about assessment items (see Facilitator's PowerPoint on CD-ROM). Introduce the "Analyzing Assessment Items Framework" (7.1.2), and answer any broad questions as the participants look at it.

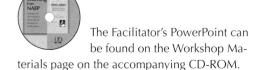
The Facilitator's PowerPoint can be found on the Workshop Materials page on the accompanying CD-ROM.

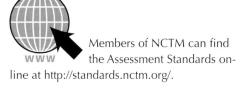
Members of NCTM can find the Assessment Standards online at http://standards.nctm.org/.

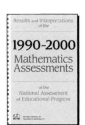
Chapter 12 of Kloosterman & Lester (2004, pp. 337–362) discusses how students' ability to construct responses relates to assessment issues. A summary of the chapter can also be found in the IU-NAEP section of the accompanying CD-ROM.

The Assessment Workshop on the CD-ROM permits the choice of a grade level and one of three options. Those options refer to specific NAEP test items and the accompanying student responses (see chart).

	4th	8th	12th
Option 1	Pizza Comparison (M045401)	Metro Rail (M070001)	Average Daily Atendance (M070601)
Option 2	Parallelogram (M041201)	Radio Station (M045901)	Center of Disk (M062401)
Option 3	Gumball Machine (M069101)		

Specific questions about the framework may arise at this point. Postpone answering very specific questions until the participants have had an opportunity to dig into the framework more deeply.

The facilitator may want to encourage participants to work with people they have not worked with recently. Also, if planning to use the items again for a different group of people, remind the participants not to write on the item cards.

 A more detailed description of this activity can be found by going to the Activity Bank in the References section of the CD-ROM and selecting the generic "Sorting NAEP Items" (Activity 5).

Analyze items

Organize the participants into groups of two or three. Give each group an item set (7.1.1) and a copy of the Item-Analysis Recording Sheet (7.1.3) for their decisions. Allow about 20 minutes for participants to analyze the items and record their results. Have participants compare analyses with someone from another group. Orchestrate a whole-group discussion about the item analysis.

The facilitator should focus the participants' attention on using the framework provided. Participants may want to interpret the language of the framework in light of previously understood meanings of words. Remind them that one goal of this workshop is to learn a shared language with which to talk about assessment items and that the authors of the framework have used their chosen words in very specific ways in the framework.

Sort items

Briefly introduce the item-sort activity, explaining the importance of thinking about mathematical items.

Organize participants into pairs. Give each couple an item set (7.1.1) and a copy of the Item-Sort Recording Sheet (7.1.4) for their decisions. Participants will sort the items into the following categories:

- Requires students to recall a memorized fact

- Requires students to perform a procedure

- Requires students to use reasoning to find an answer (a procedure may exist, but it cannot be used mindlessly)

- Requires students to use reasoning to find an answer *and* to justify their reasoning

Participants then record their decisions. Collect small-group results on an overhead transparency of the Item-Sort Recording Sheet (7.1.4). As groups finish their item sorts, collect their compiled results. If others are still working, participants can work on an item or two from the set of items (the facilitator should strategically choose an item that may have been difficult to categorize).

Put the compilation transparency on the overhead projector, and allow a minute or two for participants to look at it. The compiled results will very likely show differences in how participants categorized individual items. Ask participants for an overall reaction to the results. Then facilitate a discussion about two or three items that participants have categorized differently, putting each item on the overhead transparency as it is discussed. Ask such questions as these:

- Why did you categorize Item X as Y?

- Who agrees with this group's choice of category? What can you add to their argument?

- Who disagrees with this group's choice of category?

Many times, groups may change their minds as they listen to other groups' justifications. The focus of the discussion is not on "right" and "wrong" categorizations but on getting participants to justify categorizing the item.

Write reflectively

Have participants respond to the following prompt: Write three things you learned from this activity, two things you still have questions about, and one thing you will implement in your classroom (7.1.5, or use Facilitator PowerPoint).

ACTIVITY 2: Judge the Quality of Student Responses

The facilitator engages the participants in doing an item, looking at student work from that item, sorting the student work, creating rubrics, and then analyzing the student work using the provided NAEP rubric.

Introduction

Take a few minutes to present an overview of this portion of the workshop, making sure to emphasize that this portion of the workshop focuses on judging the quality of student work. The facilitator may want to review this section in the facilitator background notes and PowerPoint overview.

Do the NAEP item

Make an overhead transparency or PowerPoint slide of the item option (7.2.1) from the opening dialog box (pop-up menu) that offered the selection of a grade level and option. Allow participants time to work on the item alone or in pairs. The facilitator may want to participate in some of the problem solving and discussion that take place as participants work on the item.

Encourage participants to solve each item using at least two solution paths. When the participants have completed the work for an item, engage them in small-group discussions focused on the answers that the participants found for the item.

RESEARCH

"The benefits of a task-sorting activity ... accrue not simply from completing the sort, but rather from a combination of small- and large-group discussions that provide the opportunity for conversation that moves back and forth between specific tasks and the characteristics of each category ... and negotiating definitions for the categories. We have found that participants do not always agree with each other—or with us—on how tasks should be categorized, but that both agreement and disagreement can be productive" (Stein, Smith, Henningsen, & Silver, 2000).

A good follow-up to any written reflection activity is to orchestrate whole-group sharing of what the participants wrote. Many times, this whole-group sharing encourages participants to make more connections than they had individually (much like a good whole-group summary at the end of a mathematics lesson). An effective prompt to begin this discussion may be something like "Does anyone want to share with the larger group anything they wrote?"

 To choose a different option at this point, go back to the Workshop Overview page on the CD-ROM, select the Assessment Workshop, and choose a different option in the dialog box.

RESEARCH

A good resource that addresses rubrics and assessment in detail is *Developing Judgment: Assessing Children's Work in Mathematics* (Moon, 1997).

Bring at least one piece of chart paper for every group to write their rubric on, and remember to bring the tape needed to display participants' rubrics.

In this activity, most participants focus on creating a holistic rubric. After the activity is over, the facilitator may want to discuss types of rubrics that might better fit the participants' needs.

• What does the item have the potential to reveal about students' understanding of the topic?

• What mathematics is important in this item? What sorts of content-strand thinking are involved in the item?

• If a student were explaining how to do this problem, what should a teacher expect to hear or read in a good oral or written explanation? What student errors or misconceptions might be likely?

Sort student work and create rubrics

After doing the item, engage the participants in a discussion about the purpose and major types of rubrics (e.g., holistic, analytic, and descriptive). However, do not show participants a rubric, because they have the tendency to copy the rubric they have seen.

Pass out a student work packet (7.2.2) to each small group. Have participants sort the student work into piles in a way that differentiates the items. Some participants may sort the student work into only two piles, for example, "correct" and "incorrect." As a facilitator, encourage groups to think about how the work can be sorted further. Walk around the room, and try to make sense of how your participants are sorting the work to make sure that pertinent discussions emerge in the large group and small groups.

After completing the sorting activity, have the small groups discuss how they sorted the work. The conversations should focus on the decisions that were made to determine categories as well as how the student work was sorted into those categories. Ask, "What categories did you create?" "Why?" On the basis of their categorizations, have each group create its own rubric to score the work, then write the rubric on chart paper.

Gallery walk

After creating the rubrics, have participant groups hang their rubrics around the room and invite all participants to view one another's rubrics. Then have the entire group discuss how decisions were made to categorize each student response and devise the rubrics. Ask participants to discuss the qualities of the student responses that affected their decisions in creating their rubric. In addition, discuss the differences among the rubrics and why these differences may have occurred.

At the end of the Gallery Walk, the facilitator may want to discuss the importance of rubric creation in the classroom and the crucial importance of rubrics to the assessment of student learning.

Examine and use NAEP rubrics

In small groups, have the participants examine the NAEP rubric and NAEP-selected student samples (7.2.3). After examining the NAEP rubric and responses, engage participants in a large-group discussion. The following questions can serve to guide the discussion:

- Does the rubric make sense?

- How does the rubric appear to define quality?

- Does the NAEP-selected student work fit into the rubric categories?

Sort student work

Once again, pass out the student work packets (7.2.2) to each small group. This time, have participants sort the student work into the categories specified by the NAEP rubric, then compile results on the overhead comparison sheet (7.2.4). Discuss any discrepancies that may arise on the comparison sheet. The differences may lead to a discussion about the difficulty of using the NAEP rubric. At this time, consider sharing how students actually performed on this item (7.2.5; 7.2.6).

Reflective writing

In this type of reflective writing activity (7.2.7 or on the Facilitator PowerPoint), participants write for 5 minutes without stopping, using one of these prompts:

- What do you want to make sure you remember about rubrics in light of today's activities?

- What would you tell a colleague who could not be here today?

Activity 3: Gather Evidence
Introduce item choice

Explain that this activity requires homework. Participants will be choosing an item, then administering the item to their students. Have participants choose a NAEP item from the first item set (7.1.1), Activity 1: "Examine Assessment Items," or the item used in Activity 2: "Judge the Quality of Student Responses" (7.2.1).

This part of the Assessment Workshop is meant to be conducted over the course of two meetings so that the participants can administer a NAEP item to their students, create a rubric, and score the students' work between meetings.

Use the same student work packets (7.2.2) from the initial sorting activity for the "Sort student work" step.

Additional writing prompts are available in the Activity Bank on the CD-ROM. Go to the Resources section of the Activity Bank, and choose "Engaging in Reflective Writing" (Activity 14).

If the workshop is running late, the steps titled "Examine and use NAEP rubrics" and "Sort student work" can be skipped, but make sure to share with participants the NAEP rubric (7.2.3) as well as the student performance data (7.2.5) and the extended results (7.2.6) on the item chosen before doing the final reflective writing activity.

"Some participants did not have access to students to [whom they could administer the NAEP item], so we asked them to choose an item that had student work available on the CD-ROM. Before choosing their item, they checked in the NAEP Item Search on the CD-ROM to see whether it was available. Then they downloaded the student work as if it were their own. This gave them 30 student responses to score using the rubric they created."

Do the NAEP item with students

Participants should administer the item to their students and collect student work. Remind participants that they should not "teach" their students how to do the item. Then participants should create a rubric and score the student work.

Lead large-group discussion

After participants have had a chance to administer the item and score the responses using the rubric they created (give them at least a day), lead a discussion about issues related to implementing the item in the participants' classrooms. Did any surprises occur? How did the participants feel about their students' understanding of, and answers to, the problem?

Write reflectively

Have participants write for 5 minutes without stopping, using the following prompt (7.3.1):

> *On the basis of the evidence you gathered, what would be your next instructional decision?*

Share with small groups

Participants share their processes:

- Planning for assessment (choosing the NAEP item)

- Gathering evidence (collecting student work in their classrooms)

- Interpreting evidence (creating rubric; scoring student work)

- Using results (instructional "next steps")

Have participants take a moment to reflect on the entire workshop and how it may have changed the way they view and use assessment. If desired, guide the discussion by asking such questions as these:

- In what ways has this workshop influenced the ways you will think about assessment in the classroom?

- How might what you have learned in this workshop change your classroom practice?

- What insights about NAEP and standardized testing in general have you gained as a result of this workshop?

REFERENCES

Arbaugh, F., Brown, C., Lynch, K., & McGraw, R. (2004). Students' ability to construct responses (1992–2000): Findings from short and extended constructed-response items. In P. Kloosterman & F. K. Lester, Jr. (Eds.), *Results and interpretations of the 1990 through 2000 mathematics assessments of the National Assessment of Education Progress* (pp. 337–362). Reston, VA: National Council of Teachers of Mathematics.

Black, P., & Wiliam, D. (1998). Assessment and classroom learning. *Assessment in Education, 5,* 7–74.

Cooney, T. J., Badger, E., & Wilson, M. R. (1993). Assessment, understanding mathematics, and distinguishing visions from mirages. In N. L. Webb & A. F. Coxford (Eds.), *Assessment in the mathematics classroom,* 1993 Yearbook (pp. 239–247). Reston, VA: National Council of Teachers of Mathematics.

Moon, J. (1997). *Developing judgment: Assessing children's work in mathematics.* Portsmouth, NH: Heinemann.

National Council of Teachers of Mathematics (NCTM). (1995). *Assessment standards for school mathematics.* Reston, VA: NCTM.

National Council of Teachers of Mathematics (NCTM). (2000). *Principles and standards for school mathematics.* Reston, VA: NCTM.

Pellegrino, J. W., Chudowsky, N., & Glaser, R. (Eds). (2001). *Knowing what students know: The science and design of educational assessment.* Washington, DC: National Academy Press.

Senk, S., Beckmann, C.E., & Thompson, D. (1997). Assessment and grading in high school mathematics classrooms. *Journal for Research in Mathematics Education, 28,* 187–215.

Stein, M. K., Smith, M. S., Henningsen, M., & Silver, E. A. (2000). *Implementing standards-based mathematics instruction: A casebook for professional development.* New York: Teachers College Press.

Thompson, D., Beckmann, C. E., & Senk, S. (January 1997). Improving classroom tests as a means of improving assessment. *Mathematics Teacher, 90,* 58–64.

Weiss, I. R., Pasley, J. D., Smith, P. S., Banilower, E. R., & Heck, D. J. (2003). *Looking inside the classroom: A study of K–12 mathematics and science education in the United States.* Chapel Hill, NC: Horizon Research.

Catherine A. Brown and Myoungwhon Jung

THE INTENT of this workshop is to give participants an opportunity to explore both summary and item-level data on student performance for a selected state. Participants learn about the features of State NAEP and the role it plays in accountability for student learning. Much of the workshop activity is done through examining NAEP data for a selected state or other jurisdiction for which data are available.

Workshop Overview

In Activity 1, the facilitator presents a brief overview of the National Assessment of Educational Progress (NAEP) and the importance of State NAEP and asks participants to reflect on their own experiences with NAEP. Then in Activity 2, pairs of participants examine the Nation's Report Card State Mathematics 2003 Snapshot Reports for the particular state or jurisdiction of interest. The Snapshot Report provides participants with summary information about overall mathematics results, student achievement levels, performance of NAEP reporting groups, and other data. Pairs and then foursomes determine what *statements* of fact they can derive from the Snapshot Reports and also what *questions* the information stimulates for them.

In the next step of Activity 2, the foursomes of participants examine the fourth- and eighth-grade Gaps Reports for both the selected state and the nation. Those reports provide summary data for selected subgroups of students (male, female, white, black, Hispanic, not eligible and eligible for free or reduced-price lunch, and 90th and 10th percentile). Participants compare state and national gaps between subgroups and generate a list of *statements* they believe they can make and *questions* they have about the data. Finally, using the statements and questions from each foursome, written on chart paper and hung around the room, the entire group discusses what the examined NAEP data can tell them and what is still left unanswered.

The following Principles and Process Standards, taken from NCTM's *Principles and Standards for School Mathematics* (2000), are emphasized in this workshop:

NCTM Principles
- ☑ Equity
- ☑ Teaching
- ☐ Learning
- ☑ Assessment
- ☐ Technology
- ☑ Curriculum

NCTM Process Standards
- ☐ Problem Solving
- ☐ Reasoning and Proof
- ☐ Communication
- ☐ Connections
- ☐ Representation

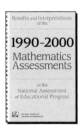

For more information about state-related issues and NAEP data, refer to chapters 8 and 15 in the monograph *Results and Interpretations of the 1990 Through 2000 Mathematics Assessments of the National Assessment of Educational Progress* (Kloosterman & Lester, 2004, pp. 193–218; 419–434). A short summary of each chapter can also be found on the CD-ROM in the IU-NAEP section.

GLOSSARY

Summary data — Information provided in the NAEP Snapshot Reports and Gaps Reports

Item-Level data — Data related to student performance on individual NAEP items

State NAEP — A state-level assessment that is identical in its content to Main NAEP but that selects separate representative samples of students for each participating jurisdiction or state because the national NAEP samples are not designed to support the reporting of accurate and representative state-level results

Snapshot Report — A one-page summary of important findings and trends in a condensed format, available for each state and jurisdiction that participated in the NAEP assessment; includes summary data about overall mathematics results, student achievement levels, performance of NAEP reporting groups, and other data

Statement — A fact or assertion offered as evidence that something is true

Question — A challenge of the accuracy, probability, or propriety of information

Gap — The difference in performance by average scale score between two groups (e.g., male and female). Gap analysis is used to determine whether all students are achieving equally in a state

Gaps Reports — Information about the gaps and changes in gaps of specific subgroups, specifically differences in gender, ethnic background, and socioeconomic status, provided to each state or jurisdiction that participates in NAEP

NAEP is the only nationally representative, continuing assessment of what United States students know and can do in various subject areas. Any state that wishes to receive a Title I grant must participate in the biennial state-level NAEP in reading and mathematics at Grades 4 and 8. State participation in NAEP other than in reading and mathematics in Grades 4 and 8 is voluntary.

In Activity 3, participants examine student performance on selected NAEP items after first doing the tasks to better understand the mathematics necessary for correctly responding to the item. They examine student performance, sample student solutions, and the scoring guide for these items, then consider implications of this information for their own curriculum and instruction.

In Activity 4, participants consolidate what they have learned and reflect on what seems most important to them. They examine summaries from chapters 8 and 15 in the monograph edited by Kloosterman & Lester (2004, pp. 193–218; 419–434). Those chapters discuss State NAEP and the relationship of NAEP to state accountability measures and student achievement. Participants then write responses to prompts that encourage them to reflect on what they have learned that was most important to them personally.

Goals

- To inform participants about NAEP, specifically the role of State NAEP as it relates to accountability for student learning

- To enable participants to interpret State NAEP summary data for their own state

- To help participants explore issues of equity by analyzing data on gaps in performance on NAEP of selected subgroups of students

- To enable participants to analyze student performance data on individual NAEP items

- To help participants develop insights into how NAEP data might influence decisions related to teaching and learning mathematics at the classroom, school, district, and state levels

Timing: 2.5 hours

Grade Band: ☑ **1–5**   ☑ **6–8**   ☐ **9–12**

NAEP Content Strand

☑ Number Sense, Properties, and Operations

☑ Algebra and Functions

☑ Data Analysis, Statistics, and Probability

☑ Geometry and Spatial Sense

☑ Measurement

 The following workshop focuses on Indiana and an item from the fourth-grade Data Analysis, Statistics, and Probability strand. To modify the workshop for other grade levels, content areas, and states, go to the Workshop section of the CD-ROM, choose State NAEP, and click on Modify.

Materials

Activity	CD-ROM No.	Materials	Number Needed
1. Overview of NAEP (25 minutes)	8.1.1	Introduction to NAEP	
2. Examine State NAEP Snapshot Report and Gaps Report (60 minutes)	8.2.1	State Mathematics Snapshot Report (by state)	1 per participant 1 overhead transparency for facilitator
	8.2.2	Gaps Report (by state)	1 per participant 1 overhead transparency for facilitator
3. Examine NAEP Test Item(s) (45 minutes)	8.3.1 8.3.2 8.3.3	Data for selected item(s) Cross-State Data (PDF) Cross-State Data (Excel)	1 per participant 1 overhead transparency for facilitator 1 per facilitator 1 per facilitator
4. Consolidate and Reflect (20 minutes)	8.4.1	Highlights from chapters 8 and 15 in Kloosterman & Lester (2004) Blank paper or large note cards	1 per participant 1 overhead transparency for facilitator 1 per participant

Background and Context Notes

NAEP is the only nationally representative, continuing assessment of what United States students know and can do in various subject areas. As a congressionally mandated project of the National Center for Education Statistics (NCES) within the U.S. Department of Education, NAEP provides a comprehensive measure of students' learning at crucial junctures in their school experience. The mathematics assessment has been conducted regularly since 1969. Because it now makes objective information about student performance available to policymakers and the general public at both the national and state levels, NAEP plays

 For convenience, cross-state data (8.3.2; 8.3.3) have been provided in two formats, PDF and Excel. Facilitators can choose the format the better suits their needs.

GLOSSARY

National Center for Educational Statistics (NCES) — The primary federal entity for collecting and analyzing data that are related to education in the United States and other nations. Under the current structure, the commissioner of education statistics, who heads the NCES in the U.S. Department of Education, is responsible by law for carrying out the NAEP project.

No Child Left Behind (NCLB) Act of 2001 — Legislation reauthorizing the Elementary and Secondary Education Act (ESEA), the main federal law affecting education from kindergarten through high school. NCLB is built on four principles: accountability for results, more choices for parents, greater local control and flexibility, and an emphasis on doing what works as verified by scientific research.

RESEARCH

State NAEP generally employs the same data-collection tools and methods as Main NAEP. Complex sampling designs are used to obtain estimates of mathematics achievement for the state as a whole as well as for population subgroups. State NAEP is not designed to provide information on individual students, teachers, schools, or school districts. Unlike Main NAEP, State NAEP is administered at Grades 4 and 8 only. (Webb & Kane, 2004)

The Web site of the U.S. Department of Education (www.ed.gov) provides official information about NCLB.

an integral role in evaluating the conditions and progress of the nation's and individual states' student education at Grades 4, 8, and 12. NAEP collects information related to academic achievement and guarantees that all data related to individual students and their families remain confidential.

NAEP has two major goals: to compare student achievement in states and other jurisdictions and to track changes in achievement of 4th, 8th, and 12th graders over time in mathematics, reading, writing, science, and other content domains. The No Child Left Behind Act of 2001 (NCLB) provides significant incentives for school districts and states to participate in NAEP.

Beginning with the 2002–2003 school year, those states that wish to receive Title I grants from the federal government *must* participate in the biennial fourth-grade and eighth-grade NAEP reading and mathematics assessments. The NAEP sample in each state is designed to be representative of the students in that state. At the state level, results are currently (2005) reported for public school students only and are broken down by several demographic subgroups of students (e.g., male, female, white, black, Hispanic, not eligible and eligible for free/reduced-price lunch, and 90th and 10th percentile). When NAEP is conducted at the state level, results are also reported for the nation.

The national NAEP sample is composed of all the state samples of public school students, as well as a national sample of non–public school students. In the event of nonparticipating states, a certain number of schools and students are selected to complete the national-level sample. For assessments conducted at the national level only, such as Long-Term Trend NAEP, samples are designed to be representative of the nation as a whole. NAEP does not, and is not designed to, report on the performance of individual students. Rather, it assesses specific populations of in-school students or subgroups of these populations, reporting on their group performance in selected academic areas. NAEP results are based on samples of these student populations of interest.

- NAEP will conduct national and state assessments at least once every two years in reading and mathematics in Grades 4 and 8. These assessments will be conducted in the same year.

- NAEP will conduct a national assessment and may conduct state assessments in reading and mathematics in Grade 12 at regularly scheduled intervals.

- To the extent that time and money allow, NAEP will be conducted in Grades 4, 8, and 12 at regularly scheduled intervals in additional subjects including writing, science, history, geography, civics, economics, foreign language, and arts.

- Any state that wishes to receive a Title I grant must include in the state plan it submits to the Secretary of Education an assurance that beginning in the 2002–2003 school year, the state will participate in the biennial state-level NAEP in reading and mathematics at Grades 4 and 8. State participation in NAEP other than reading and mathematics in Grades 4 and 8 shall be voluntary.

- State participation in other NAEP assessments is voluntary.

- No rewards or sanctions will be given to states, local education agencies, or schools on the basis of State NAEP results.

- Participation in NAEP is not a substitute for the state's own assessment of all students in Grades 4–8 in reading and mathematics.

- Local education agencies that receive a Title I subgrant must include an assurance in their Title I plans they submit to the state that they will participate in biennial State NAEP assessments of Grades 4 and 8 reading and mathematics if they are selected for the NAEP sample.

Although the No Child Left Behind legislation of 2001 directly prohibits the use of NAEP to "establish, require, or influence" state or local practices, NAEP is still viewed by many as the only available common measure of achievement across states.

Preparing to Facilitate

The facilitator (and participants, if at all possible) should read chapters 8 and 15 in the monograph edited by Kloosterman & Lester (2004, pp. 193–218; 419–434) before the workshop (summaries are available in the IU-NAEP section of the CD-ROM). Those chapters provide information regarding State NAEP and the relationship of NAEP to state accountability measures and student achievement. Although understanding the content of those chapters is not essential to the workshop, the background is helpful and supplements the context of the workshop and the background information included in the facilitator's notes. The notes that follow assume that the focus of the workshop is on data from a single state; this example uses Indiana data. The workshop can be used with Snapshot and Gaps Reports from any single state or jurisdiction that participated in State NAEP in 2003. Data for all participating states are available on the CD and can be accessed by using the Modify option for this workshop. Workshops using data from more than one state or jurisdiction are easily developed by accessing reports for the desired states or jurisdictions. In preparing for this workshop, the facilitator also needs to select one or more items to be used in Activity 3, "Examine a Set of NAEP Test Items." Twelve NAEP 2003 released items are available for use, one at each grade level for each of the five content

GLOSSARY

Sample — A portion of a population, or a subset from a set of units, that is selected by some probability mechanism for the purpose of investing the properties of the population. NAEP does not assess an entire population but rather, selects a representative sample from the group to answer assessment items.

Long-Term Trend NAEP — Recurring assessment designed to give information on the changes in the basic achievement of United States youth; administered nationally, and reports student performance at ages 9, 13, and 17 in mathematics and reading; does not evolve on the basis of changes in curricula or in educational practices

 Almost everything that one could possibly want to know about NAEP is at http://nces. ed.gov/nationsreportcard/. Much of the information in this workshop is taken directly from the Frequently Asked Questions page of this Web site. To get there, go to the Site Map and select FAQs from the About NAEP column.

RESEARCH

Analysis of State NAEP results suggests that states are far from the NCLB goal of 100 percent of students at or above the proficient level by 2014, with some states having less than 10 percent of some population subgroups at this level. During the 1990s, the percentage of students scoring below basic decreased in many states, but increases in the percentage of students attaining at or above the proficient level were smaller, especially for black and Hispanic students. (Webb & Kane, 2004)

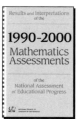

Chapter 1 of Kloosterman and Lester (2004, pp. 3–32) gives an overview of the history of NAEP, and chapters 8 and 15 (pp. 193–218; 419–434) provide information about State NAEP and its relationship to state assessments.

Participants often have little knowledge of the NAEP assessment. The PowerPoint presentation included with these workshop materials is intended to provide the background and a brief history of NAEP and State NAEP that participants often need and want. The facilitator may need to limit the discussion inspired by the presentation, however, to permit completion of the rest of the workshop in the time allotted.

strands and two items that were given at both Grades 4 and 8 (mixed), one from the Algebra strand and one from the Number Sense strand. The example presented here uses the item from the fourth-grade Data strand and materials related to it.

Facilitating the Workshop

Activity 1: Discuss History of NAEP
"Introduction to NAEP" PowerPoint

Use the "Introduction to NAEP" PowerPoint (8.1.1) to briefly lay out the history of NAEP and State NAEP. This overview will help build a context for the NAEP summary data and individual item data that participants will analyze in the workshop. The PowerPoint presentation included on the CD can be edited to fit the facilitator's needs and can be merged with the Facilitator's PowerPoint for this workshop.

Discuss NAEP and State NAEP

Ask participants what additional questions they have about NAEP and State NAEP. The background notes and chapters 1, 8, and 15 of the monograph edited by Kloosterman & Lester (2004) provide answers to some questions. The extent to which participants can engage in this discussion may depend on whether they read monograph chapters 8 and 15 beforehand.

Ask the participants about their personal experiences with NAEP. Depending on the size of the group and their comfort with one another, questions can be answered in small groups or using a facilitating strategy, such as "think, pair, share." The following questions to ask are located on the Facilitator's PowerPoint:

- What was your initial introduction to NAEP as an assessment and to NAEP data?

- How has your state performed on NAEP?

- Have your students been part of the state or national sample?

- What was your introduction to the No Child Left Behind Act?

- What information would you like about either NAEP or NCLB?

At this point, do not attempt to answer all questions, and avoid allowing the discussion to become simply complaining about federal meddling or the problems with NCLB. The purpose is to focus participants' attention on understanding the official mission of NAEP and State NAEP and how the assessment is structured. The next activity is intended to engage participants in analyzing State NAEP data from their own state.

Activity 2: Examine State NAEP Mathematics Data

Review State Snapshot Report

Pass out the State Snapshot Reports (8.2.1) for the targeted state, and ask participants to work in pairs to review the data for the targeted state and to write down one to three *statements* they can make about student performance and one or two *questions* they have. Encourage participants to make statements that they can support with the data before them and questions that the data raise for them. They should be able to point to data in the Snapshot Report that support their statements.

Have two pairs of participants come together as a foursome to discuss their statements and questions. If the foursome can answer the question using information from the report, it should become a statement; otherwise, it remains a question. If members of the foursome do not agree that a statement about student performance is supported by the Snapshot Reports, the group may decide to transform that statement into a question. After the group agrees what should be a statement and what should be a question, foursomes should write those outcomes on chart paper, placing the statements at the top and the questions at the bottom.

Examine Gaps Report

Pass out the Gaps Reports (8.2.2) for the targeted state to participants in their foursomes. Each participant should have a report for the fourth-grade and eighth-grade gaps. The facilitator may also want to have the Gaps Reports for the national sample at both grade levels available for reference (available on the CD-ROM). Participants should examine the Gaps Reports for any relationship to the statements or questions they wrote, particularly those regarding performance of different subgroups of students (e.g., black, Hispanic, white, eligibility for free or reduced-price lunch, etc.). Again, if the additional information in the Gaps Reports suggests to them that they should change a question to a statement or a statement to a question, they should indicate that change on their chart paper. If additional statements or questions emerge for the groups, they should add them to the chart paper. Then groups will tape their papers to the wall for all to view.

Discuss statements and questions

Have participants spend 5 to 10 minutes examining the posted statements and questions before asking which statements and questions participants have relating to the following considerations (also found on the Facilitator's PowerPoint):

Go to the Resources section on the CD-ROM for State Snapshot Reports and Gaps Reports for additional states as well as for the National Gaps Report for facilitator reference.

GLOSSARY

Scale score (Average scale score) — Numerical value derived from overall level of performance of groups of students on NAEP assessment items, expressed on a scale of 0 to 500. When used in conjunction with interpretive aids, such as item maps, NAEP scale scores provide information about what a particular aggregate of students in the population knows and can do.

Performance data — Information that reveals the percentage of students at each score level

---

NAEP achievement levels — Performance standards—basic, proficient, and advanced— that measure what students should know and be able to do at each grade assessed by NAEP. The achievement levels are based on recommendations from panels of educators and members of the public, and provide a context for interpreting student performance on NAEP.

Chapter 8 in Kloosterman & Lester (2004, pp. 193–218) presents some interesting data on Indiana and other states that might be introduced in this discussion as examples of what additional analyses of state-level data can tell us.

Participants may make unsupported statements about the reasons behind the gaps found in performance of student subgroups. Help participants revise those statements into questions.

A related generic activity, "Doing a NAEP Item," can be found in the Activity Bank in the Resources section of the accompanying CD-ROM.

- Performance of Indiana students in 2003

- Performance of Indiana students over Indiana's participation in State NAEP

- Performance of NAEP reporting groups in Indiana

- What NAEP scale scores and performance-level data can tell us

After participants are seated again as a whole group, facilitate a group discussion of the four areas listed above. Try to focus participants' thinking on what they can reasonably agree are "true" statements, given the data they have seen.

Participants may be confused about how scale scores and performance-level data relate. Ask them to look at the section titled Student Percentage at NAEP Achievement Levels in the state's Snapshot Report for a statement of the relationship between scale scores and performance levels (8.2.1).

In general, the information available from the Snapshot Reports and the Gaps Reports suggests that Indiana students' performance is improving, but it raises questions about which student subgroups are improving and in what areas the improvement is found. Participants should be encouraged to explore the data available on the NCES Web site (http://nces.ed.gov/nationsreportcard/) to see whether they can find answers to their questions.

Activity 3: Examine NAEP Test Item(s)

Having spent considerable time analyzing summary data on the state, participants next explore information about how students performed on specific NAEP items. However, before seeing this information, participants need to be familiar with the mathematics of an item. Accordingly, prior to examining student performance data, this activity asks participants to respond to NAEP test items, discuss their responses, and consider how responses were scored for NAEP. If some participants finish more quickly than others, suggest that they try to anticipate other responses that students may have given and why.

Do the tasks

Show the NAEP item on the PowerPoint slide, and ask participants to work on the task alone or in pairs for about 15 minutes. Alternatively, pass out the item and related data (8.3.1 and 8.3.2/8.3.3) but ask participants to look only at the item for now. Encourage participants to respond to each item as though they were trying to perform well on the assessment. Remind them to consider how students might respond and to examine the item for important mathematics as they do the task.

Discuss in small groups

When participants have responded to the items, place them in groups of three or four to share their responses. If not already done, pass out the item and related data (8.3.1 and 8.3.2/8.3.3) and ask participants to also examine the information distributed about the item. As participants discuss responses and examine information available on the item (also available on the NAEP Web site), they should address the following questions (also found on the Facilitator's PowerPoint) for each item:

- What mathematical content is important for correctly responding to this item? Where are these topics addressed in your curriculum and instruction?

- What would student responses to this item help you think about as a teacher? Regarding student thinking? In your instruction? In your curriculum?

- Is this item similar to any task you have used with your students? Is it similar to any task on an assessment used by your school?

- How would you describe the performance of Indiana students on this item compared with the national performance? Do you consider that performance satisfactory or unsatisfactory? Why?

- What issues of equity or access do the data on this item raise? How are those issues being addressed by the school, district, or state?

Revise statements and questions

Ask participants to consider whether any additional statements or questions should be added to the chart paper or whether any should be revised. Allow time for the additions or revisions.

Activity 4: Consolidate and Reflect

Having examined the Snapshot and Gaps Reports and information about individual NAEP items, participants need to consolidate what they have learned and reflect on what it means to them.

Consolidation

Pass out the highlights from chapters 8 and 15 of the Kloosterman & Lester (2004) monograph (8.4.1), and remind participants that the monograph chapters contain considerable information about State NAEP and student performance on it. Have them take a few minutes to review the highlights and consider their relationship to the statements and questions generated by the group. Briefly discuss those ideas as a whole group, focusing on what participants have learned about Indiana students and their performance on NAEP.

Reflection (5 minutes)

Pass out blank paper or note cards, and have participants respond to the following reflection prompts, which also appear on the Facilitator's PowerPoint:

- The most important statement for me was_____, and it is important to me because_____.

- The most important question for me was_____, and it is important to me because_____.

REFERENCES

Martin. W. G., & Black, J. W. (2004). State and NAEP accountability measures. In P. Kloosterman & F. K. Lester, Jr. (Eds.), *Results and interpretations of the 1990 through 2000 mathematics assessments of the National Assessment of Education Progress* (pp. 419–434). Reston, VA: National Council of Teachers of Mathematics.

Webb, N. L., & Kane, J. H. (2004). State NAEP and mathematics achievement. In P. Kloosterman & F. K. Lester, Jr. (Eds.), *Results and interpretations of the 1990 through 2000 mathematics assessments of the National Assessment of Educational Progress* (pp. 193–218). Reston, VA: National Council of Teachers of Mathematics.

Rebecca McGraw and Beatriz D'Ambrosio

I<small>N THIS</small> workshop, participants consider similarities and differences in achievement, attitudes, and opportunities to learn as those factors relate to gender, race, ethnicity, and socioeconomic standing. The NAEP data used for this workshop are organized in tables and graphs, and participants can discuss issues related to interpreting and reading across those representations, as well as think about issues of equity in teaching and learning mathematics.

Workshop Overview

This workshop has two central components: a "Gallery Walk" and an investigation of state-level data. In the Gallery Walk, enlarged tables and graphs are posted on the walls and participants move in small groups from poster to poster, analyzing data and posting comments and questions for whole-group discussion. During the investigation of state-level data, participants can use the materials on the accompanying CD-ROM or the NAEP online tools to investigate equity issues in their own state and compare their state's data with national data.

> *NAEP can help us understand how achievement, attitudes, and opportunities to learn may be different across gender, racial or ethnic, and socioeconomic groups.*

Goals

- To examine NAEP data as they relates to issues of equity

- To develop knowledge of issues surrounding the interpretation of NAEP data

- To raise questions about equity in students' performance, attitudes, and opportunities to learn and to use the Data Explorer to explore those questions in participants' own states

The following Principles and Process Standards, taken from NCTM's *Principles and Standards for School Mathematics* (2000), are emphasized in this workshop:

NCTM Principles
- ☑ Equity
- ☑ Teaching
- ☐ Learning
- ☑ Assessment
- ☐ Technology
- ☐ Curriculum

NCTM Process Standards
- ☐ Problem Solving
- ☐ Reasoning and Proof
- ☐ Communication
- ☐ Connections
- ☐ Representation

Glossary

Gallery walk — An activity in which enlarged tables and graphs are posted on the walls of the room, and participants move in small groups from poster to poster, analyzing data and posting comments and questions for whole-group discussion

Workshop

To download the materials for this workshop, go to the Workshop page on the CD-ROM and click on Manual. The Equity Workshop cuts across NAEP content strands and grade bands and is not modifiable.

- To use NAEP data when making decisions
- To interpret data presented in tables and graphs

Timing: 5 hours

Grade Band: ☑ **1–5** ☑ **6–8** ☐ **9–12**

NAEP Content Strand

☑ Number Sense, Properties, and Operations

☑ Algebra and Functions

☑ Data Analysis, Statistics, and Probability

☑ Geometry and Spatial Sense

☑ Measurement

Materials

Activity	CD-ROM No.	Materials	Number Needed
1. Introduction to Equity Issues (40 minutes)	9.1.1	NAEP monograph excerpts and focus questions	1 per participant
	9.1.2	Facilitator PowerPoint presentation	1 per facilitator
2. Gallery Walk (120 minutes)	9.2.1	Enlarged copies of NAEP monograph tables and graphs	1 per facilitator
		Sticky notes	10 per participant
	9.2.2	Regular-sized copies of NAEP monograph tables and graphs	4 of each table and graph
	9.2.3	Copies of monograph text surrounding tables and graphs	4 of each table and graph
		Overhead transparencies of NAEP monograph tables and graphs (9.2.1)	1 of each table and graph
		Blank overhead transparencies	2 per small group of participants
3. Investigate NAEP State Data (120 minutes)		Computers with Internet access	At least 1 for every 2 participants
	9.3.1	Copies of instructions for investigation of NAEP state data	1 per participant

Background and Context Notes

Through the Equity Principle, NCTM calls to the community of mathematics educators, teachers, and researchers to find a way of "making the vision of the *Principles and Standards for School Mathematics* a reality for all students" (NCTM, p. 12). NAEP analyses have suggested that students across the United States are not receiving equitable opportunities to learn mathematics, and as a result, troubling gaps are found in the performance of students of low versus high socioeconomic status, of students of different language backgrounds, of students of different genders, and of students of different ethnic backgrounds. An analysis of NAEP data lets educators explore questions about different groups of students' opportunities to learn the content, to access materials and resources, to access challenging curriculum, and to be taught by highly qualified teachers, as well as investigate many other components that can influence student learning differentially. However, caution is important in the interpretation of NAEP data. In preparation for leading participants in the discussion of equity issues related to the NAEP data, we strongly recommend that facilitators read chapters 10 and 11 in the monograph, or review the summaries of those chapters available on the CD-ROM.

Discussions of group differences, more often than not, result in characterizing members of the group as deficient in some way, for example, lacking in ability or lacking relevant cultural experiences; this view is termed the *deficit model.* According to Benjamin (1996), the deficit model is "the tendency to see characteristics of the dominant group (Whites) as the norm around which other groups vary, and against which the latter are invariably judged inferior." Through a deficit model, the blame for student underachievement is attributed to the students themselves and their families' conditions rather than to the school, curriculum, and pedagogy to which they are exposed daily and the sociocultural setting in which schools are embedded. According to Rothstein (1995), schools can be either dominating or liberating. A view of schools as liberating envisions the possibilities of working with families and communities to empower students of all socioeconomic groups, cultures, languages, and gender to succeed. "When schools dominate, consciously or not, they prepare students for the roles in society that people of their kind have historically occupied. As a liberating force, schools, consciously or not, can prepare students to break those dominating patterns and empower them to take on new roles that people from their social condition had not filled" (Rothstein, 1995, p. 46). Liberating schools identify the language and culture of all students as powerful resources that can support their learning of mathematics (Civil, 2002; Civil & Andrade, 2002).

The facilitator has a very difficult task to shift participants away from a view of differences through the lens of a deficit model toward a view of

RESEACH

"Risk is always involved in addressing bias and discrimination, but the risk for our schools and our society of not doing so is even greater" (Weissglass, 1997, p. 78).

GLOSSARY

Gap — The difference in performance by average scale score between two groups (e.g., males and females). Gap analysis is used to determine whether all students are achieving equally in a state.

Deficit model — The tendency to see characteristics of the dominant group (whites) as the norm around which other groups vary and against which the latter are invariably judged inferior

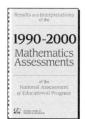

For more information about equity issues and NAEP data, refer to chapters 10, 11, and 14 in the monograph *Results and Interpretations of the 1990 Through 2000 Mathematics Assessments of the National Assessment of Educational Progress* (Kloosterman & Lester, 2004, pp. 269–304; 305–336; 383–418). A short summary of each chapter can also be found on the CD-ROM in the IU-NAEP section.

RESEACH

Extensive research uncovers complex mathematical thinking and reasoning in children and adults who have failed at school mathematics (Carraher, Carraher, & Schlieman, 1987; Saxe, 1988; Knijnik, 2002).

 Activity 2, "Using the NAEP Online Tools," from the Activity Bank in the Resources section of the CD-ROM provides a step-by-step tutorial to help participants learn how to use the three NAEP online tools.

differences through a lens that celebrates diversity and acknowledges the possibility of empowering underachieving students (Weissglass, 1997, 2001). Although most educators can embrace diversity in discussing the power of children's ideas when given opportunities for creativity, they less often celebrate diversity when they expect students to think in particular ways, such as in students' performance on standardized measures of achievement. Unfortunately, we have overemphasized the value we place on thinking in particular ways—as characterized by school mathematics—and have forgotten to create means of valuing children's thinking in situations where they are expected to be truly creative. As a result, we tend to be unable to celebrate the diversity among the students in our classrooms.

An important consideration to keep in mind is what NAEP measures. The fact that some students do not succeed on NAEP is not evidence that those students cannot think or do mathematics but only that they are struggling with school mathematics. Extensive research uncovers complex mathematical thinking and reasoning in students and adults who have failed at school mathematics (Carraher, Carraher, & Schlieman, 1987; Saxe, 1988; Knijnik, 2002).

Preparing to Facilitate

Before beginning, review chapters 10, 11, and 14 of Kloosterman & Lester (2004), the facilitator background materials (9.1.2 on the CD-ROM), and the notes in the Facilitator's PowerPoint presentation (9.1.3). To view those notes, open the PowerPoint presentation and then click on View and Notes Pages. In addition, become familiar with issues related to NAEP data interpretation; chapter 2 in this manual and Appendix B are useful introductions to the topic. If unfamiliar with using the NAEP online tools, consult chapter 3 for information about accessing state-level data via the NAEP Web site (www.nces.ed.gov/nationsreportcard). We also recommend that facilitators familiarize themselves with the graphic displays that participants will generate using the NAEP Data Explorer.

Post enlarged copies (11″ × 17″ or larger, if possible) of monograph tables and graphs on the walls around the room. Tables and graphs have been taken from chapters 10 and 11 of Kloosterman and Lester (2004) and from the *Nation's Report Card* (U.S. Department of Education, 2005) and can be downloaded from the CD-ROM (9.2.1). The tables and graphs were selected with the following purposes in mind:

- To highlight gender, racial, ethnic, and socioeconomic differences associated with achievement, attitudes, and opportunities to learn mathematics

- To consider how teachers' instructional practices are related to student achievement

- To provide examples of students' understanding of multiple content strands

- To assist participants in looking across grade levels, within grade levels, and across reporting years

- To include a wide variety of data representations

Use poster-sized copies of tables and graphs if possible for easier viewing. The tables should be posted in the following pairs: Figures 10.7 and 10.8; Figure 10.9 and Table 10.7; Tables 10.8 and 11.9.

Table Example 9.1.
Example of a NAEP monograph table (Strutchens et al., 2004, p. 272)

Average Scale Score and Percentage of White, Black, and Hispanic Students Classified at or Above the Proficient Achievement Level in 2000

		Average Scale Score	*Percent at or Above "Proficient"*
Grade 4	Nation	228	26
	White	236	34
	Black	205	5
	Hispanic	212	10
Grade 8	Nation	275	27
	White	286	35
	Black	247	6
	Hispanic	253	10
Grade 12	Nation	301	17
	White	308	20
	Black	274	3
	Hispanic	283	4

Facilitating the Workshop

Activity 1: Introduction to Equity Issues
Pre-readings for participants

Ask participants to prepare for the workshop by reading selected pages from Kloosterman & Lester (2004, pp. 383–401, 410–412). Those pages are downloadable from the CD-ROM (9.1.1).

Examine the issues

Take a few minutes to explain the purpose and format of the workshop. Next use the Facilitator's PowerPoint presentation and notes pages on the CD-ROM (9.1.2) to introduce participants to NAEP; to equity-related issues in mathematics achievement, attitude, and opportunities to learn; and to issues related to interpreting NAEP data. Finally, engage

If participants want more information about NAEP, a PowerPoint presentation reviewing the history and purpose of NAEP is available on the CD-ROM at the top of each workshop. In addition, Activity 1, "Getting an Overview of NAEP," from the Activity Bank in the Resources section of the CD-ROM provides a step-by-step guide to the PowerPoint as well as a discussion guide.

GLOSSARY

Correlation — A measure of the extent of the relation between two or more variables

Cause and effect — A measure of the extent to which changes in the value of one variable cause the value of the other variable to change

Significant — A term used to indicate that the observed changes are not likely to be associated with sampling and measurement error but are statistically dependable population differences

Sample size — The number of elements in the obtained sample

A generic version of the "Taking the Gallery Walk" (Activity 4) can be found on the CD-ROM in the Activity Bank. This and other resources can be used to build a customized workshop.

participants in a few minutes of small-group and then whole-group discussion around the following focus questions:

- How are differences in performance on NAEP among black, white, and Hispanic students (or between genders) explained by the authors?

- What data that are available through NAEP are considered primary factors influencing student performance and differences in achievement? What other explanations, for which no data are available through NAEP, might be plausible?

- Given the international perspectives provided in chapter 14, what important insights about equity should be considered in regarding NAEP as an assessment tool?

Gallery Walk
Walk the gallery

Allow small groups of participants 5 to 10 minutes per station to view the tables and graphs that have been posted around the room (9.2.1). Encourage participants to discuss the data and use their sticky notes to place comments and questions on the wall beside each graph or table. Monitor progress so that participants have time for analysis but spend the least time possible waiting to move between stations.

Discuss in small groups

Ask each small group to select a table or graph for further analysis. Distribute regular-sized copies of the selected table or graph to each group along with copies of the surrounding text from the NAEP monograph (9.2.2 and 9.2.3). In addition to considering the data presented in the table or graph in light of the surrounding text, participants should discuss the comments and questions posted by their own group and other groups during the first part of the Gallery Walk. Approximately 20 minutes should be allotted for this portion of the activity. Spend a few minutes listening to the discussion of each group and assisting with data interpretation as needed. Issues that may arise during discussion include correlation versus cause and effect, whether differences are significant, variations in sample size across reporting years, representativeness of students sampled, limitations of self-reported data (as in teacher and student surveys), and differences in national findings and local conditions. Help participants become experts with respect to their chosen table or graph in terms of the information it can and cannot provide about issues of equity.

Have large group discuss small-group findings

Distribute a transparency of the selected table or graph, blank transparencies, and overhead pens to each small group. Instruct participants to prepare short presentations (5–10 minutes) of their findings and lingering questions. For tables and graphs that create much discussion, you may want to ask the following questions:

- How are the equity issues represented in these data related to educational outcomes and career opportunities?

- What roles can participants play in the struggle to achieve equity in both opportunities and outcomes?

Activity 3: Investigate NAEP State Data
Use the NAEP Data Explorer

Lead participants in accessing the NAEP Data Explorer on the Internet (www.nces.ed.gov/nationsreportcard). Explain that the goal of the demonstration is to show methods for accessing data and creating tables and graphs. After the demonstration, participants will have time to investigate issues of interest to them. Ask them to click on Analyze Data, followed by the green Go to Advanced button. After agreeing to the terms and conditions, participants can choose a subject, grade level, state, and set of categories. Mention that only fourth-grade and eighth-grade data are available for state-level investigation. For demonstration purposes, ask everyone to choose Mathematics, an appropriate grade level and state, and, from the Major Reporting Groups category, All Students, Gender, National School Lunch Program Eligibility, and Race/Ethnicity used in NAEP reports after 2001. Finally, click on the Go to Results button.

Looking at the tables, participants can compare average scale scores across reporting years and see those data broken down by gender, race/ethnicity, and socioeconomic status. Scroll down to see the average scale scores for each subgroup for each reporting year. To create a graph of some of the data, click on "graphic" and then select the desired variable, year(s), and options. To compare across subgroup category and reporting years, select a variable, then choose "Select All" under years, "Generate graph for each jurisdiction" under Option, and "Full graph" under Graphic Options (see Figure 9.1). Note: In some states, the sample sizes of one or more racial or ethnic subgroup may be too small to permit reliable estimates of average scale scores.

Encourage participants to list three to four main points and one to two lingering questions on their blank transparencies.

According to the NAEP Web site, the NAEP online tools are designed to work on Internet Explorer and Netscape browsers, although Internet Explorer is recommended. Our experience indicates that both work well on a PC platform, but Netscape is the only browser that makes the NAEP Data Explorer fully functional on the Macintosh platform. In addition, we have found that the graphic data display in the Data Explorer works less well with a Macintosh platform.

A computer projected onto an overhead screen, if available, is very helpful for demonstration purposes.

The NAEP online tools (www.nces.ed.gov/nationsreportcard) can be used to create graphic data displays if the user's computer has an SVG (Scalable Vector Graphics) viewer installed. This software can be downloaded at no cost from www.adobe.com/svg.

If an Internet connection is not available, the state data resources can be found on the CD-ROM in the Resources section. State Snapshots and Gaps reports are available for every state.

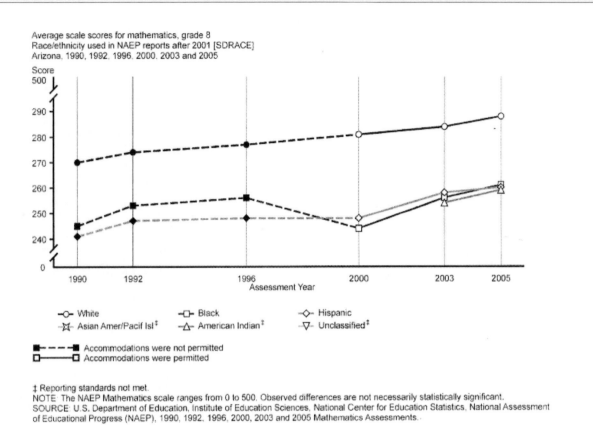

Figure 9.1. Example of Web-based NAEP data-investigation graph.

GLOSSARY

Scale score (Average scale score) — A numerical value, expressed on a scale of 0 to 500, derived from overall level of performance of groups of students on NAEP assessment items. When used in conjunction with interpretive aids, such as item maps, NAEP scale scores provide information about what a particular aggregate of students in the population knows and can do.

 Instructions for participants to use during the individual and pair investigation can be downloaded from the CD-ROM (9.3.1).

Investigate topics

Allow each participant to choose an area of interest from among the following topics:

- Instructional content and practice (e.g., courses offered, instructional tools and methods)

- Student factors (e.g., attitudes toward, and beliefs about, mathematics; mathematics courses taken)

- Factors beyond school (e.g., time spent on homework, resources in the home and community)

Participants can work individually or with a partner to investigate their areas of interest. We suggest that participants work in pairs to facilitate analysis and discussion and to help the facilitator manage the activity. Distribute copies of instructions to each participant (9.3.1). Encourage participants to begin by using the instructions to access data and create graphics, but allow time for individual explorations as well. Some of the national data analyzed during the Gallery Walk can also be compared with state-level data.

Share findings

Ask participants to share interesting findings with the whole group. If computer-screen projection equipment is available, the graphics under discussion can be created and displayed. As with the Gallery Walk, an important goal of this activity is to consider how the data relate to educational outcomes and opportunities and what roles participants can play in the struggle for equity.

Wrap up

The facilitator may want to conclude by discussing what can and cannot be learned from analyses of NAEP data. Drawing on what participants have learned from the workshop, discuss their opinions about the appropriate roles of schools in building an equitable society.

- What can be done to create equity in opportunities?

- What can be done to create equity in outcomes?

- How can the knowledge gained through a workshop such as this be shared with others and used to influence policymakers?

Pass out note cards, and ask participants to list three things they learned from this workshop, two things they plan to use from this workshop, and one thing they would like to know more about. Collect the cards from participants, or ask them to leave the cards on their tables.

REFERENCES

Benjamin, M. (1996). *Cultural diversity, educational equity, and the transformation of higher education: Group profiles as a guide to policy and programming.* Westport, CT: Praeger.

Carraher, T. N., Carraher, D. W., & Schliemann, A. D. (1987). Written and oral mathematics. *Journal for Research in Mathematics Education, 18,* 83–97.

Civil, M. (2002). Culture and mathematics: A community approach. *Journal of Intercultural Studies, 23,* 133–148.

Civil, M., and Andrade, R. (2002). Transitions between home and school mathematics: Rays of hope amidst the passing clouds. In G. de Abreu, A. J. Bishop, & N. C. Presmeg (Eds.), *Transitions between contexts of mathematical practices* (pp. 149–169). Dordrecht: Kluwer.

Kloosterman, P., and Lester, F. K., Jr. (Eds.) (2004). *Results and interpretations of the 1990 through 2000 mathematics assessments of the National Assessment of Educational Progress.* Reston, VA: National Council of Teachers of Mathematics.

Knijnik, G. (January 2002). Ethnomathematics, culture, and politics of knowledge in mathematics education. *For the Learning of Mathematics, 22,* 11–15.

Lubienski, S. T, McGraw, R., & Strutchens, M. (2004). NAEP findings regarding gender: Mathematics achievement, student affect, and learning practices. In P. Kloosterman & F. K. Lester, Jr. (Eds.), *Results and interpretations of the 1990 through 2000 mathematics assessments of the National Assessment of Educational Progress* (pp. 305–336). Reston, VA: National Council of Teachers of Mathematics.

Rothstein, S. W. (Ed.). (1995). *Class, culture, and race in American schools: A handbook.* Westport, CT: Greenwood Press.

Saxe, G. B. (August/September 1988). Candy selling and math learning. *Educational Researcher, 17,* 14–21.

Strutchens, M. E., Lubienski, S., McGraw, R., & Westbrook, S. K. (2004). NAEP findings regarding race and ethnicity: Students' performance, school experiences, attitudes and beliefs, and family influences. In P. Kloosterman & F. K. Lester, Jr. (Eds.), *Results and interpretations of the 1990 through 2000 mathematics assessments of the National Assessment of Education Progress* (pp. 269–304). Reston, VA: National Council of Teachers of Mathematics.

Sztajn, P., Anthony, H. G., Chae, J., Erbas, A. K., Hembree, D., Keum, J., Klerlein, J. T., & Tunç-Pekkan, Z. (2004). NAEP, TIMSS, and PISA: What can we learn? In P. Kloosterman & F. K. Lester, Jr. (Eds.), *Results and interpretations of the 1990 through 2000 mathematics assessments of the National Assessment of Educational Progress* (pp. 383–418). Reston, VA: National Council of Teachers of Mathematics.

Weissglass, J. (April 1997). Deepening our dialogue about equity. *Educational Leadership, 54,* 78–81.

Weissglass, J. (March 2001). Infusing equity into reform. *Leadership, 30,* 34–37.

Chapter 10
Creating Your Own Workshop: Curiosity, Curriculum, and Collaboration

Signe E. Kastberg

WORKSHOP facilitators know that nothing can compare with the thrill of learning something new with participants or colleagues. The professional energy that originates with new thoughts and feelings about teaching and learning mathematics is transformative, especially when the learner has gotten exactly what he or she needs from the experience. Although most of the workshops in this manual are modifiable, facilitators sometimes need to further personalize a workshop to meet their participants' needs. In this chapter I discuss the elements of workshop development and reflect on how I used them to meet the specific needs of participants.

In my professional experience, students simply did not use graphs spontaneously to represent or investigate data. I wondered what research in mathematics education revealed about this problem and whether other educators had had the same experience. Did activities or teaching approaches exist that support the emergence of the use of graphs to investigate data?

Curiosity

Workshop development, like lesson development, is built on the shared curiosity of learners. Facilitators and participants come together in the workshop setting to resolve their curiosities. The development of the workshop included in this chapter began with my own curiosity about graphing and the needs of preservice elementary school teachers to better understand how their students might use graphs. In my professional experience, students simply did not use graphs spontaneously to represent or investigate data. I wondered what research in mathematics education revealed about this problem and whether other educators had had the same experience. Did activities or teaching approaches exist that support the emergence of the use of graphs to investigate data?

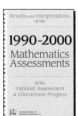

For more information about how NAEP data can help answer questions about student understanding, read the companion book to this manual, *Results and Interpretations of the 1990 Through 2000 Mathematics Assessments of the National Assessment of Educational Progress* (Kloosterman & Lester, 2004). A short summary of each chapter can also be found on the *Learning From NAEP* CD-ROM in the IU-NAEP section.

For a full listing of reference material used in the workshops, visit References in the Resources section of the accompanying CD-ROM. The user can view the references by chapter.

 Visit the NAEP Web site (http://nces.ed.gov/nationsreportcard/) to see the NAEP online tools: the NAEP Questions Tool, the NAEP Data Explorer, and the State Profiles Tool. Chapters 2 and 3 in this manual introduce the types of data available on the Web site and furnish a step-by-step guide to accessing those data.

 The user can search by content strand, grade level, or item number in the NAEP Item Search on the accompanying CD-ROM. The Item Search generates the same data set available from the NAEP online tools but expands the number of student responses to the test items to, on average, 30 responses per item.

 A Facilitator's PowerPoint presentation is available for each workshop on the *Learning From NAEP* CD-ROM. Any of these PowerPoints can be adapted to create a customized PowerPoint for a new workshop.

Facilitator

To explore these questions, I began by investigating the research literature on students' use of graphs. This search lead me to several definitive sources that focused on graph sense (Curcio, 1987; Friel, Curcio, & Bright, 2001). My colleagues and I then used these and other research findings to make sense of students' approaches to NAEP items that include graphs (D'Ambrosio, Kastberg, McDermott, & Saada, 2004).

Using the NAEP online tools on the NAEP Web site and the Item Search on the *Learning From NAEP* CD-ROM, I was able to identify graphing items and to explore fourth-grade students' performance. This pursuit led me in a new direction. I wondered how other educators thought about graphs and graph sense. How did they use and approach graphs? How did they make sense of students' work on graphing items? These questions about the sense making of colleagues drove the creation of a new workshop, an opportunity to explore my own questions and those of colleagues.

When building a customized workshop, the facilitator may find these questions helpful:

- What am I curious about?

- Why does this idea interest me?

- What sources of information can I drawn on?

Participants

Facilitators and participants have much in common; in particular, both are learners. Subjects that interest participants may be different from those that interest the facilitator, but in both instances curiosity drives learning. To build a workshop, the facilitator may need to make some assumptions about the educators who will participate in the workshop. Although areas about which educators are curious may vary a great deal, all educators want to enhance the success of their own students. Curiosity about how to help students succeed is the goal of each educator's questions. Facilitators who can tap into this source of motivation can support the development of each workshop participant as a researcher of his or her own practice.

When starting to build a workshop, consider what participants might wish to learn:

- How can I find out what interests my participants?

- What opportunities can I provide that will support participants' investigations of their own questions?

Curriculum

The design of opportunities for participants is influenced by many factors including time, materials, and facilities. Often facilitators have little or no control over those factors but must use them to plan experiences. In contrast, facilitators have more control over the activities used to engage participants in gathering information. Appendix A of this manual contains a list of NAEP-related activities that facilitators can use alone or in combination to create a workshop.

Defined broadly, curriculum includes tasks and activities as well as implementation strategies. Each of the activities in the Activity Bank on the CD-ROM includes descriptions as well as directions on engaging participants; however, each facilitator should reflect on how to select and modify activities to suit the needs of the participants. Begin by asking yourself how you can collect evidence that will help participants explore their questions and will help you explore your questions.

For example, I wondered what my participants might know or think about organizing their data. After reading the summaries in Appendix A of this manual, I went to the Activity Bank in the Resources section of the *Learning From NAEP* CD-ROM. Activity 13, "Doing the Focus Task," seemed to have been designed to elicit evidence of participants' content knowledge while providing opportunities for developing that content knowledge. Because I was interested in young children's understanding of graphing, I selected fourth-grade Data Analysis in the pop-up menu. That choice modified the activity so that I was able to view a focus task that was specific to my needs. To find out more about the activity, I downloaded the Facilitators' notes accompanying the activity. Those notes highlighted several issues of importance in selecting mathematics tasks for participants. I used the suggested task and facilitator notes to create my own focus task.

The first draft of my workshop included a focus task based on the number of pockets on each participant's clothing (Burns & Tank, 1988). Although children's knowledge of graphing and measures of central tendency may be just emerging, adults have very sophisticated ways of dealing with data. In a pilot test of the workshop using their pockets data, most participants characterized the pockets data collected from their peers by either calculating a mean number of pockets or creating a bar graph. Some discussion ensued about how adding data points might change the mean number of pockets or the graphs, but discussion was limited. The participants did not seem curious about their own data. I gathered very little evidence of the participants' understanding of graphs.

That information helped me design focus tasks that had greater potential for success. To use participants' pockets data to support questions about graphs, such as appropriate labeling of axes, I modified the task to include questions about how the connecting cube "trains" could be organized to form a graph and how the axes might be labeled. When I

 A list of fourteen generic activities is available in the Resources section of the CD-ROM Activity Bank. An overview of each activity is also available in Appendix A of this manual. Clicking on the desired activity on the CD-ROM allows the user to modify it by grade level and content strand.

Sometimes the space and resources allotted for a workshop can determine how the curriculum is implemented. For example, in two pilot tests of the workshop in this chapter, I used sidewalk chalk and large expanses of concrete and large poster boards as the "canvas" for participants' graphs. If access to a computer lab had been provided and time had been limited, I could have asked participants to use a graphing tool on a data-management software package.

 For a PowerPoint tutorial on the history and purpose of NAEP, as well as for a discussion guide, visit the Activity Bank in the Resources section of the CD-ROM (Activity 1: "Getting an Overview of NAEP").

The focus-task activity emerged from the needs of facilitators to provide participants with tasks that were challenging for adults but similar to NAEP test items. The NAEP tasks designed to elicit fourth-grade students' approaches to problems were too simplistic to engage adults and allow a facilitator to collect evidence of participants' thinking in a content area.

If participants are interested in state-specific data, go to the Resources section of the *Learning From NAEP* CD-ROM, access the State NAEP Data tab, and click on the desired state name for a downloadable State Snapshot or State Gaps Report.

 When choosing an activity from the Activity Bank in the Resources section of the CD-ROM, the user is asked to modify the activity by grade level and content area.

GLOSSARY

Evidence — Data from student work, including writing, representations, equations, drawings, and computations, that can be used to make conjectures

Conjecture — A hypothesis or assertion about what a student does or does not understand, based on the evidence in her or his work; a guess, supported by evidence, about what a student is thinking

pilot-tested the workshop including this modification, participants were engaged in lively discussion and debate.

As I developed the rest of the workshop, I kept in mind that participants want and need to explore their own questions. I returned to the Activity Bank on the CD-ROM to search for activities that would allow my participants to explore and develop conjectures about students' thinking. I knew I wanted to focus on fourth-grade students' graphing sense, so I modified each activity by grade level (4th grade) and content area (Data Analysis). I selected Activity 6, "Doing the NAEP Item," and Activity 11, "Investigating Student Work." Those activities allowed participants to explore a NAEP item and associated student work. I also modified "Doing a NAEP Item" by choosing to explore only how participants arrived at their solutions to the item and the approaches that participants thought students might take.

The design of a workshop curriculum is very personal. I used children's literature to introduce the topic of pockets and a task adapted from Burns & Tank (1988) to try to jump-start conversation. I also designed a novel task, guided by notes from Activity 13, "Doing the Focus Task," to allow participants to explore mathematics. Exploring the CD-ROM, I identified other activities that would afford opportunities to explore students' thinking. When building curriculum, a facilitator may find these questions helpful:

- What evidence will help me make conjectures about participants' thinking?

- What opportunities will allow participants to explore their own questions?

Make It Your Own

The following section is the final version of the workshop I created. It follows the same format as the other workshops in this manual, in particular, chapter 6, "Developing Mathematical Content Knowledge," but it is customized to a specific grade level and content strand and personalized to meet my needs and those of my participants'. It also incorporates a wider range of materials (e.g., literature) because of my own interests. I have included it here and on the CD-ROM to demonstrate how a facilitator can customize each of the workshops in chapters 5 through 9 of the manual by using the activities and resources available on the CD-ROM.

Exploring Graph Sense

Signe E. Kastberg

Workshop Overview

The goal of this workshop is the development of participants' content knowledge of data analysis and representation. In addition, participants will become familiar with various components of graph sense (Friel, Curcio, & Bright, 2001) through the examination of student responses to NAEP items. The workshop emphasizes representations of quantity and the importance of graph sense in students' reasoning with data.

> *Student responses are shared to facilitate partici-*
> *pants' understanding. Assumptions should not be*
> *made about students or the conditions under which*
> *they learned. Rather, the focus of the discussion*
> *should be the mathematical thinking of the students*
> *and the development of hypotheses about the stu-*
> *dents' graph sense.*

In the first activity, "Do the Focus Task," adults reason with data. Next, in "Do the NAEP Item," participants investigate how the use of pockets data can build understanding of quantity and how data can be represented and interpreted. The last activity, "Investigate Student Work," helps build an understanding of students' emerging graph sense with an eye for counting strategies, dimensions of graph having been accounted for, difficulty reading across graphs (Friel, Curcio, & Bright, 2001), and the role of context in reading graphs (D'Ambrosio et al. 2004).

 The following Principles and Process Standards, taken from NCTM's *Principles and Standards for School Mathematics* (2000), are emphasized in this workshop:

NCTM Principles

- ☐ Equity
- ☑ Teaching
- ☑ Learning
- ☑ Assessment
- ☐ Technology
- ☐ Curriculum

NCTM Process Standards

- ☐ Problem Solving
- ☐ Reasoning and Proof
- ☑ Communication
- ☑ Connections
- ☑ Representation

Workshop

This workshop is an example of how to personalize a workshop for a specific audience, grade level, and content area and therefore cannot be modified. To download materials for this workshop, go to the Workshop page on the CD-ROM and click on Manual for the workshop "Exploring Graph Sense."

Goals

- Development of content knowledge: number and operation and data analysis

- Development of links among number, data analysis, and representation

- Development of familiarity with components of graph sense

Timing: 4 hours or 2-hour sessions

Grade Band: ☑ **1–5**   ☐ **6–8**   ☐ **9–12**

NAEP Content Strand

☐ Number Sense, Properties, and Operations

☐ Algebra and Functions

☑ Data Analysis, Statistics, and Probability

☐ Geometry and Spatial Sense

☐ Measurement

Materials

Activity	CD-ROM No.	Materials	Number Needed
1. Do the Focus Task (2 hours)		One copy of *A Pocket for Corduroy* (Freedman, 1978)	1 per facilitator
		Connecting cubes	200
	10.1.1	Pockets Data (PDF)	1 per group
	10.1.2	Pockets Data (Excel)	1 per facilitator
	10.1.3	Pocket Problem (Overhead transparency)	1 per facilitator
		Poster board, graph paper, overhead transparencies, permanent markers, lined paper, rulers, unlined paper	1 set per group
	10.1.4	Prediction matrix	1 per participant
2. Do the NAEP Item (30 minutes)	10.2.1	Copies of item M049001	1 per group
3. Investigating Student Work (90 minutes)	10.3.1	Copies of student work from M049001	1 per participant
		Copies or transparencies of each piece of student work (10.3.1)	1 per facilitator, 1 set per group

Background and Context Notes

In their articulation of factors that influence students' "graph sense," Friel, Curcio, & Bright (2001, p. 132) suggest that mathematical knowledge in grades K–5 may have an impact on graph comprehension, that is, "graph readers' abilities to derive meaning from graphs created by others or by themselves." In particular the authors suggest that children's emerging understanding of number in grades K–5 may affect how they make sense of graphs. To build their understanding of numbers and foster the crucial factor in that development—unitizing (Fosnot & Dolk, 2001)—participants must investigate mathematical thought as an extension of everyday activity and problems that are age appropriate for children but possess sufficient richness to engage adults.

Friel, Curcio, & Bright also suggest that children's reasoning based on their personal experiences affects their graph sense. The authors assert that those personal experiences are often very powerful sense-making tools used to interpret contexts described in written work that accompanies graphs. Students who have experience with pockets, for example, may draw on those experiences to interpret graphs depicting data about pockets.

Facilitate the Workshop

Activity 1: Do the Focus Task
Links to literature

To provide context, the workshop begins with the facilitator reading aloud the story *A Pocket for Corduroy.* Although many issues are involved in the story, the focus in the workshop is Corduroy's quest for a pocket. This context provides a backdrop for investigating a data set involving pockets. In a mathematical extension of this activity based on Burns & Tank's pocket investigation (1988), participants develop an understanding of number and operation, especially the important concept of unitizing.

After the book is read, begin the discussion by asking why Corduroy desired a pocket of his own.

Collect and represent data

Have participants count the number of pockets they have in the clothes they are wearing. Ask participants to use the cubes to make cube trains that represent the number of pockets they have. Put the trains in a central location. Take a few minutes to discuss how the trains can be used to represent the collected data. Develop the representations that participants suggest, and ask which representation they prefer and why.

For convenience the Pockets Data set (10.1.1; 10.1.2) is offered in both PDF and Excel formats. A facilitator does not need to download both materials; the Excel file has been made available so that the data set can be modified if desired.

Part of my rationale for selecting a piece of literature to introduce our work was to hear the experiences of the participants; however, if a copy of the book is not available, the facilitator can summarize the story and invite participants to discuss Corduroy's quest: Lisa has brought her favorite teddy bear, Corduroy, along to the Laundromat. As the bear watches Lisa empty her pockets in preparation for washing, he decides that he, too, wants a pocket. The rest of the book describes his search for an appropriate pocket. The story ends with Lisa making a "pocket for Corduroy."

My first draft for the graph-sense workshop included a data-collection-and-analysis activity based on the number of pockets of each participant. In my evaluation of this task in light of reflecting on the complexity necessary to engage adult learners in mathematical thinking, I modified my original task. Participants still collected data about the number of their pockets, but I included a more complex data set to extend the activity and provide opportunities for more sophisticated data analysis and graphing strategies.

In the development of my workshop, I knew that I would have 4 hours to work with participants in a room without computer access. In addition, we had access to a long, wide hallway outside the classroom. Large expanses of concrete walkways were available outside the building where the workshop was to be held. These factors lead me to implement activities that used the resources available. For example, in one iteration of the workshop, participants drew their graphs outside using sidewalk chalk. In another iteration of the workshop, participants used poster boards and hung their graphs in the hallway for a gallery walk. I used Activity 4, Taking a Gallery Walk, in the Activity Bank on the CD-ROM to fine-tune my approach.

 To construct a customized glossary for the workshop, go to the Resources section of the CD-ROM and search the glossary for the needed words. The full glossary can be downloaded in Word format and edited as needed.

Investigate the data

Although the pockets data from a group of adults can be used to investigate issues of graph sense, adult learners often require a more challenging task to become engaged with ideas and issues of data analysis and graph sense. Pass out the Pockets Data set (10.1.1). To modify the data set, download the Excel version of this material (10.1.2). Point out that the data included with this step are more complex than the bivariate data from the pockets survey taken in the previous step. Participants are invited to investigate the data to answer a question for a clothing manufacturer (10.1.3).

> A clothing manufacturer is interested in mass-producing a new outfit but is unsure how many pockets people prefer in their outfits. To investigate the question, the manufacturer collects data during a free public event on the campus of a large urban university. Your task is to use the data to investigate and develop an answer to the manufacturer's question. Your answer must be supported by your analysis of the data. You should develop representations designed to convince the manufacturer that your findings are conclusive.
>
> Write your answer to the manufacturer's question on the back of the poster board. This question is meant to generate representations of the data that retain the richness of the original data while making clear the quantity of pockets an outfit should have.

Read and interpret graphs

Ask groups to post their representations, and allow the participants to silently interpret the graphs. Participants are asked to predict (see Prediction Matrix 10.1.4) the number of pockets that each group will suggest to answer the clothing manufacturer's question. Each participant will also be asked to identify strengths and weaknesses of the representations.

After recording their predictions, groups are asked to reveal their answers to the manufacturer's question. This phase should provoke discussions of findings, especially if differences occur between participants' predictions and the answers that the representations were designed to support or justify. Topics of discussion may include the measures of central tendency and variation. Participants will arrive at a variety of measures because they choose to eliminate various data points. The facilitator could ask how a new data point in the form of a number of pockets might change a measure of central tendency: If data from a man with twelve pockets is added to the existing data, how might the mean, median, and mode be affected?

Presentations provide the opportunity to examine the display of data. The displays can be classified as L graphs, T graphs, or pie-shaped graphs. Also discuss the specifiers, labels, and background of each graph. Reading the graphs and exploring other questions that they might be used to answer can lead to interesting discussions.

Activity 2: Do the NAEP Item
Complete the task

Distribute blank copies of the NAEP task (10.2.1) to each participant, and ask them to complete the task. The "Facilitator's Notes" in the Activity Bank for "Doing the NAEP Item" (Activity 6) can serve as a guide for this portion of the workshop.

Discuss the task

The time line for "Doing the NAEP Item" can be abbreviated, since the focus of discussion is a single task. Special attention should be paid to issues of content, number, and data analysis.

- How might students use their understanding of number?

- What graphing issues might become important?

Students may attend to only one dimension of the graph. Students and participants may bring their own understanding of pockets to the item and use that experience to make sense of the activity.

Activity 3: Investigating student work
Identify trends

Participants are given all or selected examples of students' work and are asked to identify trends or to sort the data into categories (10.3.1). Predictions based on evidence and expertise can be made about what the students know and can do. If time is short, each group can examine three pieces of student work and identify the one response they find the most interesting. They then develop questions that they would ask the child about his or her work.

Discuss as a group

These categories are then described and shared with the whole group. Students often have difficulty understanding representations that they have created as a class, despite having contributed the data themselves. An example of this difficulty, identified by Cobb (1999), is illustrated in a video resource (TERC, 2002). The Facilitator's Notes in the Activity Bank for Activity 11, "Investigating Student Work," should be used as a guideline for this portion of the workshop.

GLOSSARY

L graphs — Data displays with traditional horizontal and vertical axes

T graphs — Such data displays as tables or stem-and-leaf plots

Pie-shaped graph — Data represented as percentages of a circle, or pie. The sum of the percentages of the sectors must equal 100 percent.

Specifiers — Symbols used to represent data

 Every activity in the Activity Bank is accompanied by "Facilitator's Notes," which provide a step-by-step guide to the activity.

A good place to break the workshop into two parts is between Activities 1 and 2. In the next two activities, the participants are provided with opportunities to understand how students make sense of data and representations of data using their experience and number sense.

RESEARCH

Timmerman's (2002) results using the task and students' responses may help the facilitator prepare for participants' analysis of the task and predictions of knowledge necessary to complete the task.

GLOSSARY

Collaboration — The act of working jointly

"I have interrupted participant conversation in favor of pressing forward with workshop activities. Unfortunately, my interruptions have led to loss of momentum and in some cases, dissociation of participants from future activities."

Throughout the workshop, the facilitator should listen carefully to the tone of the comments, taking care to remind participants about issues of respect. Student responses are shared to facilitate participants' understanding. Assumptions should not be made about students or the conditions under which they learned. Rather, the focus of the discussion should be the mathematical thinking of the students and the development of hypotheses about the students' graph sense. Any opportunity to highlight issues pertinent to graph sense should be capitalized on.

Collaboration

Although good curriculum is an essential factor in workshop design, success depends largely on interactions. The work that participants engage in on a daily basis often leaves little time for collaboration. Workshops can present a forum for discussion of social, emotional, and academic issues. For facilitators, supporting conversations and collaborations provides an opportunity to learn about, and from, participants.

The success of collaborations is often based on norms that the facilitator and participants set. Sharing information without fear of ridicule is essential in fostering collaboration. Supporting the development of a safe environment is not easy. Conversations that allow the facilitator and other participants to get to know each participant and what he or she is curious about may help. Such conversations are opportunities to collect evidence that can be used during the workshop to support collaboration and push the conversation. Findings from those conversations, in which participants explore their own ideas, can also be used after the workshop to revise one's curriculum or approach.

Negotiating conversations with participants can be a challenge. Often tension arises between achieving goals set by the facilitator and allowing the needs and questions of the participants to drive the workshop.

Participants as researchers often need to pursue a line of thinking or engage in a discussion that a facilitator might not see as productive. Fervent exploration of the activity or discussion in which the group is engaged is necessary for the good health of collaboration. Although each situation is different, I tend to support the development of collaborations in the directions of interest to participants, even at the cost of the goals of *my* workshop. I justify this approach by remembering two points:

- my goals may not be those of the participants, and

- I may not understand why the conversation is important.

Finally, my reflections following the workshop allow me to understand the critical nature of the conversations and make adjustments to the workshop to support such conversations in the future.

REFERENCES

Burns, M., & Tank, B. (1988). *A collection of math lessons from grades 1 through 3.* Sausalito, CA.: Math Solutions.

Cobb, P. (1999). Individual and collective mathematical development: The case of statistical data analysis. *Mathematical Thinking and Learning, 1*(1), 4–43.

Curcio, F. R. (1987). Comprehension of mathematical relationships expressed in graphs. *Journal for Research in Mathematics Education, 18,* 382–393.

D'Ambrosio, B., Kastberg, S. E., Mermott, G., and Saada, N. (2004). Reading and interpreting data in disciplines other than mathematics. In P. Kloosterman & F. K. Lester, Jr. (Eds.), *Results and interpretations of the 1990 through 2000 mathematics assessments of the National Assessment of Education Progress* (pp. 363–381). Reston, VA: National Council of Teachers of Mathematics.

Fosnot, C., & Dolk, M. (2001). *Young mathematicians at work: Constructing number sense, addition, and subtraction.* Portsmouth, NH: Heinemann.

Freedman, D. (1978). *A pocket for Corduroy.* New York: Viking Press.

Friel, S. N., Curcio, F. R., & Bright, G. W. (2001). Making sense of graphs: Critical factors influencing comprehension and instructional implications. *Journal for Research in Mathematics Education, 32* (2), 124–158.

TERC. (2002). *Developing mathematical ideas: Working with data.* Parsippany, NJ: Dale Seymour.

Timmerman, M. A. (2002). Learning to teach: Prospective teachers' evaluation of students' written responses on a 1992 NAEP graphing task. *School Science and Mathematics, 102,* 346–358.

Appendixes

Appendix A: Activity Bank

1. Getting an Overview of NAEP

2. Using the NAEP Online Tools

3. Understanding Data

4. Taking the Gallery Walk

5. Sorting NAEP Items

6. Doing a NAEP Item

7. Developing Rubrics

8. Examining Rubrics

9. Investigating Extended Multiple-Choice Assessment

10. Examining Distractors

11. Investigating Student Work

12. Understanding Evidence, Conjectures, and Curricular Implications

13. Doing the Focus Task

14. Engaging in Reflective Writing

Appendix B: NAEP Data Primer

1. NAEP Sampling

2. NAEP Assessment Administration

3. NAEP Item-Level Data

4. NAEP Scale Score Data

5. Inferences from the NAEP Scale Score Data

Appendix C: Glossary of Terms

Activity 1: Getting an Overview of NAEP

Activity Description

In this activity, participants learn about the genesis and development of the National Assessment of Educational Progress (NAEP). Following a modifiable PowerPoint presentation, participants explore the purpose and structure of NAEP and examine the impact of NAEP data on current testing practice. In small groups, participants discuss how NAEP has affected their own educational perspectives and consider the impact of NAEP on local and national educational policy.

Goals

- To inform participants about NAEP and the role of NAEP as it relates to accountability and student learning

- To help participants develop insights into how NAEP data might be used to influence educational policy locally and nationally

Timing: 45 minutes

Grade Band: ☑ 1–5   ☑ 6–8   ☑ 9–12

NAEP Content Strand

☑ Number Sense, Properties, and Operations

☑ Algebra and Functions

☑ Data Analysis, Statistics, and Probability

☑ Geometry and Spatial Sense

☑ Measurement

Materials

1.1 Facilitator Background Materials

1.2 NAEP Overview PowerPoint slide

1.3 Questions for Discussion overhead transparency

Activity 2: Using the NAEP Online Tools

Activity Description

The NAEP online tools (NAEP Questions Tool, NAEP Data Explorer, and State Profile Tool) allow anyone having interest in NAEP to easily access important NAEP data, including released NAEP items, a comparison of the NAEP results on the basis of demographic and background characteristics, and NAEP data for individual states. This activity is designed to help participants make effective use of the NAEP online tools. In this activity, the facilitator takes a step-by-step approach to help participants learn how to use three NAEP online tools. As an extension, participants learn how and where to find information that interests them by engaging in some of the tasks.

Goals

*   To provide easy access to all NAEP data for individual states and for the whole nation

*   To help participants understand different features of each NAEP online tool

*   To help participants make effective use of the NAEP online tools to find information that interests them

Timing: 90 minutes

Grade Band: ☑ 1–5   ☑ 6–8   ☑ 9–12

NAEP Content Strand

☑ Number Sense, Properties, and Operations

☑ Algebra and Functions

☑ Data Analysis, Statistics, and Probability

☑ Geometry and Spatial Sense

☑ Measurement

Materials

2.1 Facilitator Background Materials
    Computer with projector and Internet access

2.2 Summary of Chapter 2 (PowerPoint slides)

2.3 Task sheets for each participant

2.4 Facilitator copy of answers for each task

Activity 3: Understanding Data

Activity Description

This activity is designed to give participants hands-on experience with common data-analysis procedures used with classroom and NAEP data. Participants work tasks that examine central tendency and significance within NAEP and classroom data. Appendix B of this manual provides a short primer on those concepts and works in concert with this activity.

Goals

- Teachers will gain understanding of NAEP data by completing tasks using data from classroom contexts

- Teachers will gain experience with the NAEP online data tools

- Teachers will be exposed to common statistical terms used in reporting NAEP data

Timing: 60 minutes

Grade Band: ☑ 1–5   ☑ 6–8   ☑ 9–12

NAEP Content Strand

☐ Number Sense, Properties, and Operations

☐ Algebra and Functions

☑ Data Analysis, Statistics, and Probability

☐ Geometry and Spatial Sense

☐ Measurement

Materials

3.1 Facilitator Background Materials

3.2 Part I Worksheet

3.3 Part II Worksheet

Activity 4: Taking the Gallery Walk

Activity Description

Participants analyze tables and graphs of NAEP data. Figures and tables of NAEP data are posted on the walls, and small groups of participants move from item to item, analyzing data and posting comments and questions for whole-group discussion. After all participants have visited each item, small groups choose one item of interest to further analyze. The groups use copies of the table or figure, the text that surrounds it in the monograph, and other reference materials (e.g., copies of the descriptions of NAEP achievement levels, descriptions of NAEP question types) to respond to various questions and comments posted by the groups. Finally, small groups share their findings and any remaining questions with the other groups.

Goals

- To support participants in developing knowledge of issues surrounding the interpretation of NAEP data

- To help participants use NAEP data as a source of information that can guide instructional decision making

- To engage participants in discussing issues related to interpreting data presented in table and figures and reading across those representations

- To give participants opportunities to develop knowledge of NAEP results related to a particular area of interest (e.g., grade band, item type, demographic subgroup)

Timing: 120 minutes

Grade Band: ☑ 1–5   ☑ 6–8   ☑ 9–12

NAEP Content Strand

☐ Number Sense, Properties, and Operations

☐ Algebra and Functions

☑ Data Analysis, Statistics, and Probability

☐ Geometry and Spatial Sense

☐ Measurement

Materials

4.1 Facilitator's Background Materials

4.2 Selected tables and graphs

Sticky notes for each participant

Enlarged copies and overhead transparencies of selected tables and graphs

Blank overhead transparencies and pens

4.3 Copies of monograph text that surrounds tables and graphs

Activity 5: Sorting NAEP Items

Activity Description

In this activity, participants work in pairs or triples to sort a set of NAEP items, using a predetermined set of criteria. After participants sort the tasks, they engage in a whole-group discussion of their results. This activity furnishes participants with a common language for analyzing mathematical tasks that they can use to evaluate textbooks, tasks used in their mathematics classes, and mathematics tasks on state and national assessments.

Goals

* To engage participants in learning a language for talking about mathematical items

* To engage participants in using that language to analyze mathematical items

* To engage participants in thinking deeply about items that their students encounter on mathematics assessments

Timing: 90 minutes

Grade Band: ☑ 1–5 ☑ 6–8 ☑ 9–12

NAEP Content Strand

☑ Number Sense, Properties, and Operations

☑ Algebra and Functions

☑ Data Analysis, Statistics, and Probability

☑ Geometry and Spatial Sense

☑ Measurement

Materials

5.1 Facilitator Background Materials

5.2 Set of tasks and overhead transparency

5.3 Recording sheet and overhead transparency

Activity 6: Doing a NAEP Item

Activity Description

In this activity, participants work individually or in pairs completing NAEP items. They work the items as open-ended items, even if the items were originally given in multiple-choice format. In small groups participants discuss their answers and consider how their students might respond to these test items. This phase gives participants firsthand experience in solving the items and may lead to a discussion of common errors. Finally, participants engage in a large-group discussion of their findings, including common errors for each NAEP item.

Goals

*   To highlight the importance of analyzing mathematical items to determine the cognitive thinking required to solve them

*   To raise participants' awareness of how NAEP items may differ in their levels of cognitive demand

Timing: 60 minutes

Grade Band: ☑ 1–5   ☑ 6–8   ☑ 9–12

NAEP Content Strand

☑ Number Sense, Properties, and Operations

☑ Algebra and Functions

☑ Data Analysis, Statistics, and Probability

☑ Geometry and Spatial Sense

☑ Measurement

Materials

6.1 Facilitator Background Materials

6.2 Facilitator Notes—Monograph Data

6.3 Open-ended NAEP Items for each participant

6.4 Questions for Discussion overhead transparency

6.5 Facilitator Notes—Correct Answers for each item

6.6 Facilitator Notes: Data Table on Student Performance

Activity 7: Developing Rubrics

Activity Description

In this activity, participants work individually or in small groups to create rubrics to judge the quality of students' work. In small groups, participants sort a set of student work in a way that makes sense to them. After the sorting process, participants discuss the sort and create a rubric that would be useful for assessing the student work.

Goals

- To think critically about a collection of student work, forming judgments about the quality by using a sorting process

- To create rubrics that reflect the way a collection of student work was sorted

Timing: 90 minutes

Grade Band: ☑ 1–5 ☑ 6–8 ☑ 9–12

NAEP Content Strand

☑ Number Sense, Properties, and Operations

☑ Algebra and Functions

☑ Data Analysis, Statistics, and Probability

☑ Geometry and Spatial Sense

☑ Measurement

Materials

7.1 Facilitator Background Materials

7.2 Hard copies of the task

7.3 Hard copies of the student work packets from the task

7.4 NAEP Rubric and Student Exemplars facilitator copy

7.5 Data on Student Performance facilitator copy

Activity 8: Examining Rubrics

Activity Description

In this activity, participants work individually or in small groups to analyze selected student work given the NAEP rubric. Participants make judgments about student work by sorting them into the NAEP rubric categories. Participants next create a master chart of how each group sorted the student work, then discuss the discrepancies. At the end of the activity, they evaluate the utility of the NAEP rubric.

Goals

- To examine the NAEP rubric for a particular task

- To think critically about a collection of student work, forming judgments about the quality by sorting the work using the NAEP rubric

- To think critically about the utility of a NAEP rubric

Timing: 90 minutes

Grade Band: ☑ 1–5   ☑ 6–8   ☑ 9–12

NAEP Content Strand

☑ Number Sense, Properties, and Operations

☑ Algebra and Functions

☑ Data Analysis, Statistics, and Probability

☑ Geometry and Spatial Sense

☐ Measurement

Materials

8.1 Facilitator Background Materials

8.2 Hard copies of task for each participant

8.3 Hard copies of the student work packets

8.4 Hard copies of the NAEP Rubric and representative samples of each category

8.5 Data on Student Performance (facilitator copy and overhead transparency)

8.6 Comparison Sheet (hard copy and overhead transparency)

Appendix A

Activity 9: Investigating Extended Multiple-Choice Assessment

Activity Description

In this activity, participants explore a different type of multiple-choice item—extended multiple choice (E-MC)—to learn how it may reveal student understanding. E-MC encourages students to go beyond simply picking an answer from among several options. In addition to selecting a correct answer, students must also indicate how certain they are about their choice, and they are instructed to X out any options that they are certain are incorrect. Partial credit is awarded according to a scheme that rewards correct answers that are not due to guessing, and also rewards the ability to detect incorrect choices.

This activity is adapted from *Dynamic Classroom Assessment* by George W. Bright and Jeane M. Joyner (2004). This activity works well when done in combination with "Doing a NAEP Item" (see Activity 6) or "Examining Distractors" (see Activity 10). If combined with another activity, the use of the same set of NAEP items is recommended. An example of a workshop combining all three activities can be found in chapter 5.

Goals

- To help teachers think about how the structure of test items affects how much students can show about what they know, and how much teachers can detect about students' thinking

- To introduce the extended-multiple-choice (E-MC) test format as a possible new alternative

- To explore how E-MC can foster teachers' insights into the thinking of their students

Timing: 90 minutes

Grade Band: ☑ 1–5   ☑ 6–8   ☑ 9–12

NAEP Content Strand
☑ Number Sense, Properties, and Operations
☑ Algebra and Functions
☑ Data Analysis, Statistics, and Probability
☑ Geometry and Spatial Sense
☑ Measurement

Materials

9.1 Facilitator Background Materials

9.2 Facilitator Notes—Monograph Data

9.3 Blank Extended Multiple-Choice Answer Sheet

9.4 Example Multiple-Choice Item overhead transparency

9.5 Instructions for Scoring Extended Multiple-Choice (E-MC) Items

9.6 Worksheet to Practice Scoring E-MC Test Items

9.7 Multiple-Choice NAEP Items

9.8 Facilitator Notes—Data Table on Student Performance

9.9 Questions for Discussion overhead transparency

Activity 10: Examining Distractors

Activity Description

In this activity, participants carefully examine the real distractors from a set of NAEP items. Working in small groups, participants attempt to predict which NAEP distractors will be popular and unpopular student responses, and discuss why they think so. As a large group, they discuss the actual percentage of respondents for each choice presented in a data table. The activity leads participants toward a decision about what makes a good multiple-choice distractor as well as how to best use multiple-choice items in the classroom. Participants become more engaged in examining the distractors if they have had the opportunity to work the test items. This activity works best when preceded by "Doing a NAEP Item" (see Activity 6).

Goals

- To raise participants' awareness of how NAEP items may differ in their levels of cognitive demand

- To inform participants about NAEP trends in content-strand instruction and student performance

- To help participants make informed decisions about when and how to best use multiple-choice assessments in the classroom

Timing: 60 minutes

Grade Band: ☑ 1–5 ☑ 6–8 ☑ 9–12

NAEP Content Strand

☑ Number Sense, Properties, and Operations

☑ Algebra and Functions

☑ Data Analysis, Statistics, and Probability

☑ Geometry and Spatial Sense

☑ Measurement

Materials

10.1 Facilitator Background Materials

10.2 Facilitator Notes—Monograph Data

10.3 Multiple-Choice NAEP Items

10.4 NAEP Distractors Data Table

10.5 Distractor Questions PowerPoint slide

Activity 11: Investigating Student Work

Activity Description

This activity is designed to help participants develop their understanding of students' mathematical thinking. In this activity, individual participants or small groups examine student work from NAEP tasks. Participants generally find the examination of student work interesting and can spend substantial amounts of time discussing issues raised. Evidence from this examination is used to make conjectures about mathematics content and processes used by the students to craft solutions. The facilitator conducts a large-group discussion of conjectures that emerge from small-group discussion.

Goals

- To help participants become aware of or better understand students' mathematical thinking

- To help participants identify and analyze the fundamental mathematics content and process used by students in their mathematics work

- To help participants develop expertise in making evidence-based conjectures about students' mathematical thinking

- To give participants opportunities to develop their mathematical knowledge

Timing: 90 minutes

Grade Band: ☑ 1–5   ☑ 6–8   ☑ 9–12

NAEP Content Strand

☑ Number Sense, Properties, and Operations

☑ Algebra and Functions

☑ Data Analysis, Statistics, and Probability

☑ Geometry and Spatial Sense

☑ Measurement

Materials

11.1   Facilitator Background Materials

11.2   Student work packet

Activity 12: Understanding Evidence, Conjectures, and Curricular Implications

Activity Description

In this activity, participants focus on the important issues brought up by a single NAEP task. First, participants sort the student work on a continuum from exemplary to unsatisfactory. Participants then select four student answers from different sections of the spectrum of work, and consider the evidence for each. They make conjectures about each student's knowledge, and recommend curricular implications for each student.

Goals

- To examine the knowledge necessary to understand a mathematical concept

- To be aware that students can produce incorrect answers and at the same time demonstrate correct ideas and thinking

- To develop ways to help children better understand a concept by looking at evidence, making conjectures about children's understanding, and discussing curricular implications for future instruction

Timing: 90 minutes

Grade Band: ☑ 1–5   ☑ 6–8   ☑ 9–12

NAEP Content Strand

☑ Number Sense, Properties, and Operations

☑ Algebra and Functions

☑ Data Analysis, Statistics, and Probability

☑ Geometry and Spatial Sense

☑ Measurement

Materials

12.1   Facilitator Background Materials

12.2   Student Responses for the NAEP Item
       Student Responses overhead transparencies for facilitator

12.3   "Moving Toward Understanding"
       Blank overhead transparencies and pens for participants to share their work

Activity 13: Doing the Focus Task

Activity Description

In this activity, participants work in small groups to complete an intriguing task that is linked by content strand with a NAEP task. The focus task is designed for participants working with students in the grade band indicated in the NAEP problem. The purpose of this task is to focus the participants' thinking on the mathematics that is addressed in the NAEP task.

Goals

- To present a task that requires thoughtful analysis on a level of thinking that is higher than that required by the NAEP task

- To focus participants' attention on the mathematical content they will encounter when they perform the NAEP task

- To cause participants to experience some of the disequilibrium and challenge that students experience when performing NAEP tasks

Timing: 30 minutes

Grade Band: ☑ 1–5   ☑ 6–8   ☑ 9–12

NAEP Content Strand

☑ Number Sense, Properties, and Operations

☑ Algebra and Functions

☑ Data Analysis, Statistics, and Probability

☑ Geometry and Spatial Sense

☑ Measurement

Materials

13.1   Facilitator Background Materials

13.2   Hard copy of the task for each participant

Overhead transparency of the task

Blank overhead transparencies and pens for participants to record their thinking

13. 3   Discussion of the Focus Task facilitator notes

Activity 14: Engaging in Reflective Writing

Activity Description

This activity contains a number of reflective writing suggestions for facilitators to use in a workshop.

Goals

- To engage teachers in reflecting on their learning

- To encourage teachers to connect what they have learned with their instructional decisions in the classroom

- To encourage teachers to connect what they have learned with their students' learning

Timing: 5 to 10 minutes

Grade Band: ☑ 1–5   ☑ 6–8   ☑ 9–12

NAEP Content Strand

☑ Number Sense, Properties, and Operations

☑ Algebra and Functions

☑ Data Analysis, Statistics, and Probability

☑ Geometry and Spatial Sense

☑ Measurement

Materials

14.1   Facilitator Background Materials

Appendix B: NAEP Data Primer

This appendix deals with the "nuts and bolts" of the NAEP mathematics assessments. In chapters 2 and 3, we explained how to access and understand the NAEP mathematics assessment data. In this appendix, we focus on the unique features of the NAEP assessment, the data, and the NAEP online tools.

NAEP Sampling

 Beginning in 2002, as a result of the redesign of the NAEP assessment in 1998, a combined sample was used to select schools for both state and national NAEP. Selecting a subset of schools from the state samples is an effort to reduce the total number of schools participating in NAEP. From the schools, a subsample is selected as the national subset.

The State NAEP population is composed of students in Grades 4 and 8 in public schools in participating states. (State NAEP does not include Grade 12. A separate sample is selected for national and regional estimates.) NAEP can be used as an indicator of student achievement across the states and perhaps the nation. If NAEP results indicate an increased gap in student achievement in a particular subject area, questions will arise (NCES, 2002).

For the NAEP results to represent the complete student population, each student subpopulation must be accurately represented. This requirement translates statistically into every student in the population having a nonzero probability of selection in a sample. However, if some students have zero probability of selection in a sample, then those students are not represented in the NAEP results (NCES, 2002). Ideally, every student has an equal opportunity of selection in a sample.

Currently, states are required to sample at least 2500 students from at least 100 schools for each subject assessed. In 1994, a policy was established that allowed fewer than 100 schools in the sample if fewer than 100 schools were eligible. However, then the sample size is the total number of eligible schools. Ideally, 25 students are assessed at each school for each subject for which the school was selected (Chromy, 2003).

The properties of a NAEP sample are very different from those of a simple random sample because of the complex sample design. Since all students in the target population do not have an equal opportunity of selection, the sample weights are used in the data analysis to account for these differences. Population and subpopulation characteristics based on the assessment data are estimates determined using the sampling weights (NCES, 2001b). The NCES uses the *jackknife method,* which is a method of estimation that adjusts sample weights so they sum to specified population totals that correspond to the levels of a particular response variable (NCES, 2003).

NAEP Assessment Administration

The NAEP assessment administration staff is a group of well-trained individuals who coordinate and administer the NAEP assessments in the selected schools. In 2003, the NAEP mathematics administration consisted of two 25-minute mathematics question sets. At the time of the assessment, NAEP staffers administer background questionnaires to students, teachers, and schools.

Background Questionnaires

Students respond to the questionnaire items on a voluntary basis. Their names and responses are kept confidential. The student questionnaire consists of two 5-minute sets of questions that immediately follow the content questions in their assessment booklets. Those students who choose to participate provide background information, such as demographic characteristics, educational experiences, and opportunities for additional support in learning either in or out of school. An alternative questionnaire is provided for limited English proficiency students and students with disabilities. This questionnaire is printed in a separate booklet and is usually completed by a teacher or staff member who knows the student.

Teachers and schools may complete their versions of the questionnaire either electronically on the NAEP Web site or on paper. The teacher background questionnaire asks teachers to respond to questions related to demographic characteristics, professional training and development, and classroom and instructional practices. The school questionnaires are usually completed by the principal or assistant principal and include information on school policies and characteristics.

NAEP sorts the data gathered from the teacher and school questionnaires into the eight categories of Major Reporting Groups, Student Factors, Factors Beyond School, Instructional Content and Practice, Teacher Factors, School Factors, Community Factors, and Government Factors. Those data are available to the public on the Nation's Report Card Web site (http://nces.ed.gov/nationsreportcard).

Item Blocks

Each student chosen to participate in the NAEP assessment receives an assessment booklet. Due to the design of the item blocks, described subsequently, several different assessment booklets are distributed to participants. No one individual student completes all the NAEP mathematics assessment items. Instead, each student receives an assessment booklet containing two 25-minute question sets.

The items used in the two question sets are carefully selected through the use of a *balanced incomplete block (BIB)* spiraling procedure. Students receive text booklets with blocks of questions. The BIB spiraling procedure assigns a block of questions to the first, second, or third position in a booklet an equal number of times. Every block is paired with every other block of questions. (Thirteen blocks of questions were used for the 2000 mathematics assessment.) The spiraling procedure cycles the test booklets such that typically only a small number of students receive the same booklet in an assessment session (NCES, 2001b).

NAEP Item-Level Data

The NAEP data available through the online NAEP Data Explorer provide row percentages for released questions. Row percentages in a table represent the number of students in an individual table cell, divided by the total number of students in the row, converted to a percentage. In other words, the row percentages represent the proportion of students answering in each response category. Figure B.1 shows the row percentages for a mathematics question on the 2003 assessment. The asterisk indicates choice C as the correct answer. Additionally, the NAEP Data Explorer allows the comparison of performance of males and females on each item. Figure B.2 shows the row percentages by gender for the same mathematics question on the 2003 assessment.

Figure B.1. Row percentage for item.

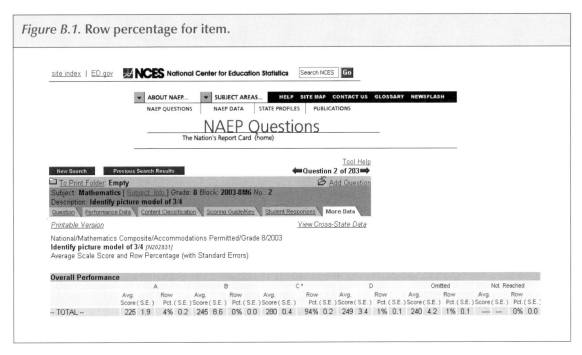

Figure B.2. Row percentage by gender.

Gender	A			B			C *			D			Omitted			Missing	
	Avg. Score (S.E.)	Row Pct.	(S.E.)	Avg. Score (S.E.)	Row Pct.	(S.E.)	Avg. Score (S.E.)	Row Pct.	(S.E.)	Avg. Score (S.E.)	Row Pct.	(S.E.)	Avg. Score (S.E.)	Row Pct.	(S.E.)	Avg. Score (S.E.)	Row Pct.
Male	224 (2.6)	5 (0.3)	‡ (‡)	# (0.1)	282 (0.5)	93 (0.3)	247 (3.9)	1 (0.1)	240 (4.0)	1 (0.1)	‡ (‡)	# (
Female	227 (2.1)	3 (0.2)	‡ (‡)	# (0.0)	279 (0.4)	95 (0.3)	252 (5.8)	1 (0.1)	239 (7.4)	1 (0.1)	‡ (‡)	# (

NAEP Scale Score Data

Once NAEP data are scored and compiled, responses are weighted according to the sample design and population structure and then adjusted for nonresponse. Weighting ensures that the students are represented in the NAEP results in the same proportion as they are in the school population in the grades assessed. State sampling weight reflects the probability that a student is selected in a sample when school and student nonresponse are taken into consideration (NCES, 2003). The weighting used in poststratification ensures that the subpopulations are represented according to the U.S. Census and the Current Population Survey. Poststratification reduces sampling errors and makes adjustments for nonresponse. NAEP uses a technique called *ranking ratio adjustments.* Final sampling weights are computed by ranking the composite weights to match known population and subpopulation totals. This raking results in adjusted weights used for estimation (NCES, 2001a).

Central Tendency

The measure of central tendency used by NAEP in reporting scale score is the mean scale score. The NAEP data available via the online NAEP Data Explorer supplies the mean, or average, scale score of a given assessment. No individual student scores are assigned to NAEP assessments. No student completes the entire NAEP assessment. Instead, students complete a variety of blocks of items. Through a fairly complicated weighting process, NAEP statistically determines the expected score for a given student. The scale score represents the expected score of a given student had the student completed the entire set of questions. Figure B.3 shows a table taken from the online NAEP Data Explorer. The average scale score of the national 4th-grade mathematics assessment in 2005 was 238. The average scale score represents the expected score of a student who completes the entire assessment.

Figure B.3. NAEP online Data Explorer average scale score.

Average scale scores for mathematics, grade 4, All students [TOTAL]: By jurisdiction, 2005

All students	Year	Jurisdictions	Average Scale Score	Standard Error
All students	2005	National	238	(0.1)

NOTE: The NAEP Mathematics scale ranges from 0 to 500. Observed differences are not necessarily statistically significant.
SOURCE: U.S. Department of Education, Institute of Education Sciences, National Center for Education Statistics, National Assessment of Educational Progress (NAEP), 2005 Mathematics Assessment.

Inferences from the NAEP Scale Score Data

Along with offering a rich set of educational data to the public, the National Center for Education Statistics offers a set of data-analysis tools for use at the Nation's Report Card Web site (NAEP Questions Tool, NAEP Data Explorer, State Profiles). Before describing the statistical tools available on the site, let us review a few basics of inferential statistics.

Variability

As we summarize distributions of data using measures of central tendency, we must keep in mind the degree of variance of the values within the distribution. The variability of a distribution describes how closely the values in the distribution lie around the mean. A small variability indicates that the values of the distribution lie close to the mean of the distribution. A large variability indicates that values of the distribution are dispersed farther away from the mean.

The standard deviation is a useful measure of the variability of a distribution. It is a value that represents the distribution's spread. The formula for the standard deviation of a distribution is

$$SD = \sqrt{\frac{\sum(X - \overline{X})^2}{n}},$$

where $\sum$ represents a sum, X represents individual values, $\overline{X}$ represents the mean of the distribution, and n represents the number of values in the distribution. In words, the standard deviation is equal to the square root of the sum of the squares of the differences of the mean and the individual distribution values divided by the number of values in the distribution.

Example:

Find the mean and standard deviation of the following distribution:

80, 100, 50, 61, 74

Solution:

First find the mean of the distribution.

$$\overline{X} = \frac{80+100+50+61+74}{5} = \frac{365}{5} = 73$$

Next find the values needed for the formula giving the standard deviation.

Deviation (Score – Mean)

Score	Mean	$X - \overline{X}$	$(X - \overline{X})^2$
80	73	7	49
100	73	27	729
50	73	−23	529
61	73	−12	144
74	73	1	1

$$\sum (X - \overline{X})^2 = 1452$$

$$S = \sqrt{\frac{\sum(X-\overline{X})^2}{n}} = \sqrt{\frac{1452}{5}} = \sqrt{290.4} = 17.04$$

Standardized Scores

Standardized scores, such as z scores, provide a way to compare individual scores across testing instruments. By converting raw scores to standardized scores, one can establish a common scale that allows for a comparison of one individual with another individual. The z score conveys the distance between a score and a mean in units of standard deviation.

Because the z distribution is a normal distribution (50% of the scores lie on each side of the mean), we can find the probability of the occurrence of a particular z score. Figure B.4 shows the percentage of scores falling within a range of z scores.

According to figure B.4, 68% percent of the scores are expected to fall within the range from −1 to 1 z scores, 95% within the range from −2 to 2 z scores, and 99.7% within the range from −3 to 3 z scores. Therefore, the probability that a particular z score falls outside the range −1 to 1 is 32% (100% − 68%). Similarly, the probability that a particular z score falls outside the range −2 to 2 is 5% (100% − 95%), and outside the range −3 to 3 is 0.3% (100% − 99.7%).

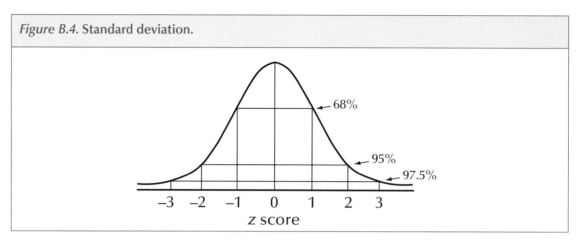

Figure B.4. Standard deviation.

Inferential Statistics

An understanding of the *z* distribution acts as a precursor for *inferential statistics*. Inferential statistics encompasses a set of statistical tests in which samples are used to make generalizations about the population. A test common to inferential statistics is a comparison of sample means. By performing a *t* test, one can determine, within a certain probability, whether two sample means are from the same population.

A *t* test allows a comparison of two means to determine whether a statistically significant difference exists between the two of them. Performing a *t* test yields a *p* value, which is the probability that the two means come from the same population. The standard level of significance is often set at 5%, or 0.05. On the one hand, if the *p* value is 0.05 or less, we can say that we are 95% confident that the means come from different populations. That is, we can say that the two means are significantly different at the 0.05 level. On the other hand, if the *p* value is greater than 0.05, we assume that the means come from the same population. That is, the two means are not significantly different at the 0.05 level.

Example:

> You collect your students' English and mathematics scores on last year's standardized achievement tests. You are interested in determining whether a significant difference exists between the 52 scores of the males and the 53 scores of the females. The English scores indicate that the average score of the males is 88 with a standard deviation of 2.00, whereas the average score of the females is 87 with a standard deviation of 2.25. The average mathematics score of the males is 92 with a standard deviation of 3.25, whereas the average mathematics score of the females is 93 with a standard deviation of 3.0.
>
> You perform a t test on the English scores of the males and females and obtain a *p* value of 0.0178. The *p* value for the mathematics scores is 0.1046. Are the English scores and mathematics scores statistically significant?

Statistical Significance

The most commonly accepted level of statistical significance is $p < 0.05$. In the foregoing example, the English scores are significantly different because the *p* value is less than 0.05. However, the mathematics scores are not significantly different because the *p* value is greater than 0.05. Notice that even though the difference in scores is only one point in both instances, the standard deviation plays a major role in significance testing. In this example, differences in the standard deviation cause a difference of 1 point to be significant in one data set and not in the other.

Just as in the previous example, NAEP scale scores are standardized scores. We can perform significance tests on subsets of the scores to determine differences between groups. Figure B.5 shows the 2005 average mathematics scale scores for males and females. The figure shows that the average scale score of the males is 239 and the standard error is 0.2; the average scale score of the females is 237 and the standard error is 0.2. To find out whether the scores are significantly different, click Find Out from the "Are differences statistically significant?" window that appears above the data table.

Figure B.5. 2003 average mathematics scale scores by gender.

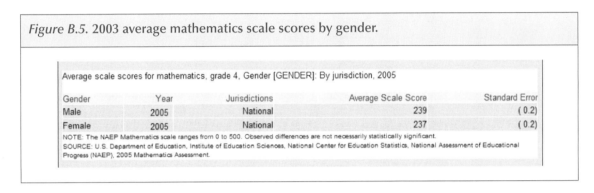

The results of the *t* test are shown in Figure B.6. The key at the bottom indicates the significance. NAEP considers *p* values of 0.05 and less as significant.

Figure B.6. NAEP online Data Explorer *t*-test results.

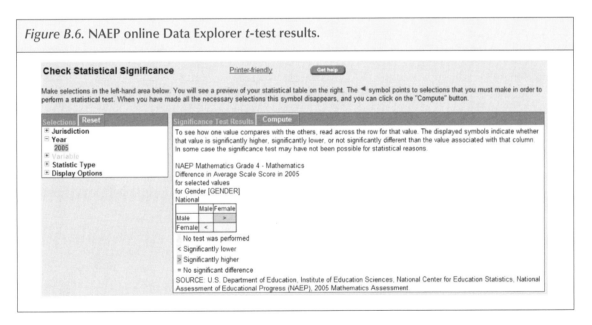

Starting in 2003, the number of students participating in the NAEP mathematics assessments grew dramatically. An important point to keep in mind is that even small differences become significant as the sample size grows larger. Therefore, the same difference in scale score may be significant in some administration years and not in others. An example of this issue occurs with the average mathematics scale scores of males and females in the years 2000 and 2003. In 2000, the average scale score for males was 227 and for females, 224—a difference of 3 scale points. In 2003, the average score for males was 236 and the average score for females was 233, again a difference of 3 scale points. The 3-point difference in 2000 is not significant; however, the 3-point difference realized in 2003 is significant. The larger sample size in 2003 plays a part in this distinction.

Additional Resources

The National Center for Education Statistics provides numerous documents aimed at disseminating NAEP data. The ones listed here are particularly interesting in light of the topics discussed in this appendix.

- *How the samples of schools and students are selected for the Main Assessments (state and national).* Available at http://nces.ed.gov/nationsreportcard/about/nathow.asp

- *How is the NAEP mathematics assessment administered?* Available at http://nces.ed.gov/nationsreportcard/mathematics/howadmin.asp

- *Background questionnaires.* Available at http://nces.ed.gov/nationsreportcard/bgquest.asp

- *2003 NAEP demonstration booklet: Grade 4.* Available at http://nces.ed.gov/nationsreportcard/pdf/demo_booklet/gr4demobook2003.pdf

- 2003 NAEP Demonstration Booklet: Grade 8. Available at http://nces.ed.gov/nationsreportcard/pdf/demo_booklet/gr8demobook2003.pdf

REFERENCES

Chromy, J. R. (2003). *The effects of finite sampling corrections on state assessment sample requirements.* Retrieved January 30, 2005, from http://nces.ed.gov/pubsearch/pubsinfo.asp?pubid=200317

National Center for Education Statistics (NCES). (2003). *Handbook of survey methods.* Retrieved January 30, 2005, from http://nces.ed.gov/pubs2003/2003603.pdf

National Center for Education Statistics (NCES). (2002). *An agenda for NAEP validity research.* Retrieved January 30, 2005, from http://nces.ed.gov/pubs2003/200307.pdf

National Center for Education Statistics (NCES). (2001a). *Technical report and data file user's manual for the 1992 national adult literacy survey.* Retrieved January 30, 2005, from http://nces.ed.gov/pubs2001/2001457.pdf

National Center for Education Statistics (NCES). (2001b). *The nation's report card: Mathematics 2000.* Retrieved January 30, 2005, from http://nces.ed.gov/nationsreportcard/pdf/main2000/2001517.pdf

Appendix C: Glossary of Terms

The definitions included in this glossary and the chapter glossaries were obtained or adapted from several different resources. Most of the terms were found in the glossary available on the NAEP Web site (NCES, 2005) or in the NAEP 1996 and 2000 frameworks (NAGB, n.d.). Additional definitions were located through the use of Internet search engines (Benjamin, 1996; Mueller, 2005; North Carolina State University, 2005).

A

Accommodations

> Alterations in the administration of standardized assessments, such as NAEP, that are provided to certain students with disabilities (SD) or limited English proficiency (LEP), as specified in the student's Individualized Education Program (IEP)

Achievement-level percentages

> Values that indicate the percentage of students within the total population, or in a particular subgroup, that meet or exceed expectations of what they should know and be able to do; specifically, the weighted percentage of students with NAEP composite scores that are equal to, or that exceed, NAGB specified achievement-level cut scores

Achievement levels

> Performance standards, based on recommendations from panels of educators and members of the public, set by the National Assessment Governing Board (NAGB) to provide a context for interpreting student performance on NAEP. The levels—basic, proficient, and advanced—measure what students should know and be able to do at each grade assessed.

Activity bank

> A section of the CD providing access to a set of generic activities that can be used as building blocks to create or modify a workshop

Advanced

> One of the three NAEP achievement levels, denoting superior performance at each grade assessed

Advanced search

> A search for questions by grade, content classification, question type, difficulty, and other subject-specific variables

Alternative

> A response choice included in a multiple-choice item; option

Analytic rubric

> A scoring scheme that articulates levels of performance for each criterion so the teacher can assess student performance on each criterion

B **Background questionnaires**

The instruments used to collect information about student demographics and educational experiences

Balanced Incomplete Block (BIB) spiraling

A complex variant of multiple matrix sampling in which items are administered so that each pair of items is dispensed to a nationally representative sample of respondents

Basic

One of the three NAEP achievement levels, denoting partial mastery of prerequisite knowledge and skills that are fundamental for proficient work at each grade assessed

Basic search

A NAEP Web site tool, existing in the online Data Explorer, that allows the user to specify the parameters of subject, grade, state/jurisdiction, and category to search for data

Block

A group of assessment items created by dividing the item pool for an age or grade into subsets; used in the implementation of the BIB spiral sample design

Booklet

A paperback assessment book, created by combining blocks of assessment items

C **Cause and effect**

A measure of the extent to which changes in the value of one variable cause the value of the other variable to change

Collaboration

The act of working jointly

Common block

A group of background items included at the beginning of every assessment booklet

Composite scale

An overall subject-area scale based on the weighted average of the scales that are used to summarize performance on the primary dimensions of the curricular framework for the subject-area assessment. For example, the mathematics composite scale is a weighted average of five content-area scales: Number Sense, Properties, and Operations; Measurement; Geometry and Spatial Sense; Data Analysis, Statistics, and Probability; and Algebra and Functions. These five scales correspond to the five content-area dimensions of the NAEP mathematics framework.

Conjecture

A hypothesis or assertion about what a student does or does not understand, based on the evidence in her or his work; a guess, supported by evidence, about what a student is thinking

Constructed-response item

A non-multiple-choice item that requires some type of written or oral response

Content classification

A designation of the mathematical content area(s) and knowledge that the selected item assesses

Content knowledge

Knowledge specific to mathematics or a particular mathematical content strand

Correlation

A measure of the extent of the relation between two or more variables

Curricular implication

Actions to be taken to further a student's understanding, based on evidence and conjectures made from examining the student's responses

Curriculum

A program of instruction; includes tasks and activities as well as implementation strategies

D

Data Explorer

An online tool offering tables of detailed results from NAEP national and state assessments using data based on information gathered from the students, teachers, and schools that participated in NAEP

Deficit model

The tendency to see characteristics of the dominant group (whites) as the norm around which other groups vary and against which the latter are invariably judged inferior

Descriptive rubric

A scoring scheme that creates broad categories, such as setting up the problem, and scores each area separately (Moon, 1997)

Disequilibrium

A state at which a person has reached the limits of his or her knowledge or has encountered a situation that does not fit into his or her conceptual framework. Further inquiry and experience can resolve this uncomfortable state by helping the person understand the problem or concept and perceive that it "makes sense."

Distractor

An incorrect response choice included in a multiple-choice item

Downloadable materials

Materials that can be transferred from the CD to a personal computer

E

Evidence

Data from student work, including writing, representations, equations, drawings, and computations, that can be used to make conjectures

Expected value

The average of the sample estimates given by an estimator across all possible samples. If the estimator is unbiased, then its expected value will equal the population value being estimated.

Extended

The highest scoring level on an extended constructed-response item, indicating that the student's response matches the description given in the NAEP scoring rubric

Extended constructed-response item

An item in which the situation demands more than a numerical or short verbal response, instead requiring the student to carefully consider a problem within or across the content strands, understand what is required to solve the problem, choose a plan of attack, carry out the attack, and interpret the solution in terms of the original problem

Extended multiple-choice (E-MC) item

A test item that requires more thought and response from students than a traditional multiple-choice item, is easily scored, awards partial credit, and can offer educators more insight into student thinking than an item in traditional multiple-choice format. E-MC can take many forms. In addition to picking an answer option from a list provided, students may be asked to indicate how sure they are about their choice, or to indicate which rejected options they know are wrong and which they are uncertain about, or to explain why the options they labeled wrong are incorrect, or to provide some other explanation of their thinking.

F

Field test

Items in NAEP mathematics and reading assessments at Grades 4 and 8 go through two levels of pretesting: pilot test and field test. A field test is the second stage of pretesting for these assessments, through which the student assessment instrument for the following year is finalized. The field test is administered to a nationally representative sample of students one year before the operational assessment. The field-test results are used to make IRT scaling decisions and perform precalibration, thereby speeding reporting.

Focus task

A task to be explored by workshop participants to immerse them in the content of a given NAEP item before they solve the NAEP item itself. Such a task is intended to help participants experience some of the struggle that a student might encounter when solving a NAEP item.

Focused BIB spiraling

A variation of BIB spiraling in which items are administered so that each pair of items within a subject area is dispensed to a nationally representative sample of respondents

Formative assessment

A feedback process that furnishes information that can be used to fine-tune or modify an existing instructional approach

Framework

An underlying structure supporting the analysis of assessment items

G

Gallery walk

An activity in which enlarged tables and graphs are posted on the walls of the room, and participants move in small groups from poster to poster, analyzing data and posting comments and questions for whole-group discussion

Gap

The difference in performance by average scale score between two groups (e.g., males and females). Gap analysis is used to determine whether all students are achieving equally in a state.

Gaps Reports

> Information about the gaps and changes in gaps of specific subgroups, specifically differences in gender, ethnic background, and socioeconomic status, provided to each state or jurisdiction that participates in NAEP

Glossary

> A list of definitions of the terms used throughout this manual and the workshops

Grade level

> Schooling level at which NAEP examinations are administered: Main NAEP at the 4th-, 8th-, and 12th-grade levels; State NAEP at the 4th- and 8th-grade levels

H

Highlights

> Bulleted points from selected chapters in the NAEP monograph

High-stakes test

> An assessment instrument used to make significant educational decisions about children, teachers, schools, or school districts

Holistic rubric

> A scoring scheme that assigns an overall level of performance by assessing performance across multiple criteria

I

Incorrect

> The lowest scoring level on an extended constructed-response item, indicating that the student demonstrates no understanding

Item

> The basic, scorable part of an assessment; a test question

Item-level data

> Data related to student performance on individual NAEP items

Item response theory (IRT)

> Test-analysis procedures that assume a mathematical model for the probability that a given examinee will respond correctly to a given exercise

J

Jackknife

> A procedure that estimates standard errors of percentages and other statistics; particularly suited to complex sample designs

K

Keyword search

> NAEP Web site tool, existing in the online Data Explorer, that allows the user to type in a keyword to search for data on specific background topics

L

L graphs

Data displays with traditional horizontal and vertical axes

Limited English proficient (LEP)

A term used to describe students who are in the process of acquiring English language skills and knowledge. Some schools refer to these students using the term *English language learners* (ELL).

Long-Term Trend NAEP

Recurring assessment designed to give information on the changes in the basic achievement of United States youth; administered nationally, and reports student performance at ages 9, 13, and 17 in mathematics and reading; does not evolve on the basis of changes in curricula or in educational practices

M

Main NAEP

An assessment instrument that reports information for the nation and specific geographic regions of the country, includes students drawn from both public and nonpublic schools, and reports results for student achievement at Grades 4, 8, and 12

Mathematical knowledge

The bank of knowledge on which a person draws when confronted with a mathematical problem

Minimal

A low scoring level on an extended constructed-response item, indicating that the student demonstrates very little understanding

More data

Additional data for each NAEP item, including the performance of various subgroups on the item

Multiple-choice item

An item that consists of one or more introductory sentences followed by a list of response options that include the correct answer and several incorrect alternatives

N

NAEP (National Assessment of Educational Progress)

The only nationally representative, continuing assessment of what United States students know and can do in various subject areas

NAEP ability levels

The three general mental abilities associated with mathematics and targeted as primary foci in NAEP assessments: conceptual understanding, procedural knowledge, and problem solving

NAEP achievement levels

Performance standards—basic, proficient, and advanced—that measure what students should know and be able to do at each grade assessed by NAEP. The achievement levels are based on recommendations from panels of educators and members of the public, and provide a context for interpreting student performance on NAEP

NAEP content strands

The five mathematics content areas used in the Main and State NAEP examinations: (a) Number Sense, Properties, and Operations, (b) Measurement, (c) Geometry and Spatial Sense, (d) Data Analysis, Statistics, and Probability, and (e) Algebra and Functions

NAEP item number

The number used to identify an individual NAEP item; typically begins with the letter M followed by six digits. Item numbers can be found on the More Data page, following the item description.

NAEP item search

A tool available on the CD-ROM that allows the user to choose additional student responses or select different NAEP items

NAEP Questions Tool

A tool available on the CD-ROM that accesses information by tabs that mirror those in the NAEP online Questions Tool on the NCES Web site (discussed in chapter 3):

— Question
— Performance Data
— Content Classification
— Scoring Guide
— Student Responses
— More Data

NAEP scales

The scales common across age or grade levels and assessment years, used to report NAEP results

NAEP Web site (online data tools)

World Wide Web site, www.nces.ed.gov/nationsreportcard/, maintained by the National Center for Educational Statistics, that has links to NAEP items, student performance, questionnaire results, and state-specific data

National Assessment Governing Board (NAGB)

An independent, bipartisan organization of individuals appointed by the U.S. Secretary of Education to give overall policy direction to the NAEP program. Its members include governors, state legislators, local and state school officials, educators, business representatives, and members of the general public.

National Center for Educational Statistics (NCES)

The primary federal entity for collecting and analyzing data that are related to education in the United States and other nations. Under the current structure, the commissioner of education statistics, who heads the NCES in the U.S. Department of Education, is responsible by law for carrying out the NAEP project.

National School Lunch Program (NSLP)

A federally assisted meal program that provides low-cost or free lunches to eligible students. It is sometimes referred to as the *free/reduced-price lunch program*. Free lunches are offered to those students whose family incomes are at or below 130 percent of the poverty level; reduced price lunches are offered to those students whose family incomes are between 130 percent and 185 percent of the poverty level.

No Child Left Behind (NCLB) Act of 2001

Legislation reauthorizing the Elementary and Secondary Education Act (ESEA), the main federal law affecting education from kindergarten through high school. NCLB is built on four principles: accountability for results, more choices for parents, greater local control and flexibility, and an emphasis on doing what works as verified by scientific research.

Nonresponse

A lack of response or measurement for a given sample element

Not reached

An item to which the student did not respond because the time limit was reached for the section of the assessment on which he or she was working. After the first "not reached" item, the student will have given no responses to any further questions on that section of the assessment.

O

Off-task response

A response that is unrelated to the question being posed; differs from an incorrect response or an omitted response

Omitted

No response given

Open-ended item

A type of assessment item in which students must construct an answer rather than simply choose from among answer options that are provided. NAEP open-ended items come in several formats: items in which students simply write their answers in the space provided (short constructed-response, or SCR), items in which students answer multiple questions and provide a brief rationale for each response given (another sort of SCR), or items in which students provide extended constructed-response (ECR) answers.

Option

Response choice included in a multiple-choice item

P

Partial

A scoring level on an extended constructed-response item, indicating that the student's explanation is only to some extent correct

Percent correct

The percentage of a target population who answered a particular exercise correctly

Percentile

A score location below which a specified percentage of the population falls. For example, in 1998, the tenth percentile of fourth-grade reading scores was 167. This statistic means that in 1998, 10 percent of fourth-graders had NAEP reading scores below 167 while 90 percent scored at or above 167.

Performance data

Information that reveals the percentage of students at each score level

Pie-shaped graph

Data represented as percentages of a circle, or pie. The sum of the percentages of the sectors must equal 100 percent.

Pilot test

A pretest of items to obtain information regarding clarity, difficulty levels, timing, feasibility, and special administrative situations. The pilot test is performed before revising and selecting the items to be used in the assessment or, in the instance of mathematics and reading at Grades 4 and 8, before selecting items to be used in the field test.

Population

A group being studied. In NAEP, the population of interest is the entire collection of United States students in public or private schools at Grades 4, 8, or 12 (or in the Long-Term-Trend Assessments, at ages 9, 13, and 17 years). The small samples of students that NAEP selects for the assessment permit inferences about academic performance to be made about all school students at the three grade or age levels.

Population mean

The average of all the items in a group of individuals (population) being sampled

Probability sample

A sample in which every element of the population has a known, nonzero probability of being selected

Proficient

One of the three NAEP achievement levels, indicating that students reaching this level have demonstrated competence over challenging subject matter, including subject-matter knowledge, application of such knowledge to real-world situations, and analytical skills appropriate to the subject matter

Public and nonpublic schools

The types of school that students attend and according to which NAEP results are reported. Nonpublic schools include Catholic and other private schools. Because they are funded by federal authorities, not state or local governments, Bureau of Indian Affairs (BIA) schools, Department of Defense Domestic Dependent Elementary and Secondary Schools (DDESS), and Department of Defense Dependents Schools (Overseas) are not included in either the public or nonpublic categories; they are included in the overall national results.

Q

Question

A challenge of the accuracy, probability, or propriety of information

Question tab

Provides the NAEP item as it is given on the assessment

Questions Tool

Online tool furnishing easy access to NAEP questions, student responses, and scoring guides that are released to the public; presents both national and state data, where appropriate

Quick search

A search for NAEP questions by subject and grade, available as a data tool on the CD-ROM in the NAEP Item Search section

R **Race**

A human subpopulation whose members share physical and genetic similarity. In all NAEP assessments, data about student race/ethnicity is collected from two sources: school records and student self-reports. Before 2002, NAEP used student self-reported race and ethnicity as the source of information reported for race/ethnicity. In 2002, the decision was made to change the student race/ethnicity variable highlighted in NAEP reports. Starting in 2002, NAEP reports of race/ethnicity are based on school records; student self-reported race/ethnicity will continue to be reported in the NAEP Data Explorer.

Raw score

The unadjusted assessment score, equivalent to the number of correct responses

Reference number

Number included in the workshop table to identify each item used in the workshop

Released item

A test question that has been made available to the public. After each assessment, NCES releases nearly one-third of the questions.

Reporting subgroups

Groups within the national population for which NAEP data are reported, for example, gender, race/ethnicity, grade, age, level of parental education, region, and type of location

Representative sample

A portion of a population, or a subset from a set of units, that is selected by some probability mechanism for the purpose of investigating the properties of the population. NAEP does not assess an entire population but rather, selects a representative sample from the group to answer assessment items.

Response options

In a multiple-choice question, alternatives that can be selected by a respondent

Resources

A section of the CD-ROM containing the activity bank, references, glossary, and Nation's Report Card information

Row percentage

In a tabular presentation (such as in the NAEP Questions Tool), the number of students represented in a particular cell of the table, divided by the number of students in the row of the table, converted to a percent

S **Sample**

A portion of a population, or a subset from a set of units, that is selected by some probability mechanism for the purpose of investing the properties of the population. NAEP does not assess an entire population but rather, selects a representative sample from the group to answer assessment items.

Sample mean

The average of all the items in a sample. If the sample is chosen carefully, the sample mean is a good estimate of the population mean.

Sample size

The number of elements in the obtained sample

Sampling error

The error in survey estimates that occurs because only a sample of the population is observed; measured by sampling standard error

Satisfactory

A scoring level on an extended constructed-response item, indicating that the student's explanation is adequate

Scale score (Average scale score)

A numerical value, expressed on a scale of 0 to 500, derived from overall level of performance of groups of students on NAEP assessment items . When used in conjunction with interpretive aids, such as item maps, NAEP scale scores provide information about what a particular aggregate of students in the population knows and can do.

School questionnaire

A questionnaire completed for each school by the principal or other official; used to gather information concerning school administration, staffing patterns, curriculum, and student services

Scoring guide/key

A list of the correct answers for multiple-choice questions; also, a holistic rubric used to score each student's response for short constructed-response and extended constructed-response items

SD/LEP questionnaire

A questionnaire completed for each selected student identified as a student with a disability (SD) and/or limited English proficiency (LEP) by the school staff member most knowledgeable about the student

Search by block

Within a subject, a feature to search for a specific block, or booklet, of questions that were administered to students who participated in the NAEP assessment

Session

A time frame during which a group of students reports for the administration of an assessment. Most schools conduct only one session, but some large schools conduct ten or more.

Short constructed-response item

A type of short-answer item that requires students to (a) give either a numerical result or the correct name or classification for a group of mathematical objects, (b) draw an example of a given concept, or (c) write a brief explanation for a given result

Significant

A term used to indicate that the observed changes are not likely to be associated with sampling and measurement error but are statistically dependable population differences

Simple random sampling

The process for selecting n sampling units from a population of N sampling units so that each sampling unit has an equal chance of being in the sample and every combination of n sampling units has the same chance of being in the sample chosen

Snapshot Report

A one-page summary of important findings and trends in a condensed format; given to each state and jurisdiction that participated in the NAEP assessment; includes summary data about overall mathematics results, student achievement levels, performance of NAEP reporting groups, and other data

Specifiers

Symbols used to represent data

Standard deviation

A measure of the dispersion of a set of scores, specifically, the square root of the average squared deviation of scores about their arithmetic mean

Standard error

A measure of sampling variability and measurement error for a statistic; estimated, because of NAEP's complex sample design, by jackknifing the samples from first-stage sample estimates; may also include a component that reflects the error of the measurement of individual scores estimated using plausible values

Standardized score

A converted raw score that offers a way to compare scores across testing instruments

State NAEP

A state-level assessment that is identical in its content to Main NAEP but that selects separate representative samples of students for each participating jurisdiction or state because the national NAEP samples are not designed to support the reporting of accurate and representative state-level results

State Profile

Online tool that presents important data about each state's student and school population and its NAEP testing history and results, providing easy access to all NAEP data for participating states and links to the most recent State Report Cards for all available subjects

Statement

A fact or assertion offered as evidence that something is true

Statistically significant

A term used to indicate that the observed changes are likely due to true differences between groups rather than sampling and measurement error

Student responses

Student answers that have been scored on the basis of the NAEP scoring rubric; not available for multiple-choice questions

Students with disabilities (SD)

Pupils with an array of physical, emotional, or learning limitations, who may need specially designed instruction to meet their learning goals. A student with a disability usually has an Individualized Education Program (IEP), which guides his or her special education instruction. Students with disabilities are often referred to as *special education students* and may be classified by their school as learning disabled (LD) or emotionally disturbed (ED).

Subgroups

Groups of the student population identified in terms of certain demographic or background characteristics. Some of the major reporting subgroups used for reporting NAEP results include students' gender, race, ethnicity, highest level of parental education, and type of school (public or nonpublic). Information gathered from NAEP background questionnaires also makes possible the reporting of results on the basis of such variables as courses taken, home discussions of schoolwork, and television-viewing habits.

Subject area

One of the areas assessed by NAEP, for example, art, civics, computer competence, geography, literature, mathematics, music, reading, science, U.S. history, or writing

Summary data

Information provided in the NAEP Snapshot Reports and Gaps Reports

Systematic sample

A sample selected by a systematic method

T **T graphs**

Such data displays as tables or stem-and-leaf plots

Teacher questionnaire

A questionnaire completed by selected teachers of sample students; used to gather information concerning years of teaching experience, frequency of assignments, use of teaching materials, and availability and use of computers

Tests of significance

Statistical tests conducted to determine whether the changes or differences between two numerical results are statistically significant. The term *significant* does not imply a judgment about the absolute magnitude or educational relevance of changes in student performance; rather, it is used to indicate that the observed changes are not likely to be associated with sampling and measurement error but are statistically dependable population differences.

V **Variance**

The average of the squared deviations of a random variable from the expected value of the variable. The variance of an estimate is the squared standard error of the estimate.

W **Weighted percentage**

A percentage that has been calculated by differently weighting cases. It differs from a simple percentage in which all cases are equally weighted.

Workshop

Professional development experiences addressing particular themes related to the analysis of NAEP data: student understanding, content knowledge, assessment, state issues, and equity. Each workshop in this manual and CD can be modified by NAEP content strand, state, or grade level.

About These Professional Development Materials

This workshop facilitator's manual and its accompanying CD are designed to help facilitators create meaningful professional development experiences for classroom teachers of mathematics and mathematics educators at all grade levels. The activities in these *Learning From NAEP* materials are designed to help educators better understand the intricacies of assessment data and how they can use such data to support student learning in the mathematics classroom. The workshops address student understanding, content knowledge, assessment, and equity, and the facilitator can modify each workshop by NAEP content strand, state, or grade level.

These professional development materials are the result of the second of two primary foci of a larger project, IU-NAEP, a collaborative effort of the National Council of Teachers of Mathematics (NCTM) and mathematics educators at Indiana University with financial support from the National Science Foundation. Focus I of the IU-NAEP project prepares interpretive reports of data from various NAEP mathematics assessments and has produced two monographs (see below): *Results and Interpretations of the 1990 Through 2000 Mathematics Assessments of the National Assessment of Educational Progress,* edited by Peter Kloosterman and Frank K. Lester Jr. (Reston, VA: NCTM, 2004), and *Results and Interpretations of the 2003 Mathematics Assessment of the National Assessment of Educational Progress,* edited by Peter Kloosterman and Frank K. Lester Jr. (Reston, VA: NCTM, forthcoming).

Also Available From NCTM . . .

Joint NCTM and IU-NAEP Focus I Interpretative Monographs:

*   *Results and Interpretations of the 1990 Through 2000 Mathematics Assessments of the National Assessment of Educational Progress,* edited by Peter Kloosterman and Frank K. Lester Jr. (Reston, VA: National Council of Teachers of Mathematics, 2004)

*   *Results and Interpretations of the 2003 Mathematics Assessment of the National Assessment of Educational Progress,* edited by Peter Kloosterman and Frank K. Lester Jr. (Reston, VA: National Council of Teachers of Mathematics, forthcoming)

Please consult www.nctm.org/catalog for the availability of these titles and for a plethora of resources for teachers of mathematics at all grade levels.

For the most up-to-date listing of NCTM resources on topics of interest to mathematics educators, as well as on membership benefits, conferences, and workshops, visit the NCTM Web site at www.nctm.org.